Real Perspective of Fourier Transforms and Current Developments in Superconductivity

Edited by Juan Manuel Velazquez Arcos

Published in London, United Kingdom

IntechOpen

Supporting open minds since 2005

Real Perspective of Fourier Transforms and Current Developments in Superconductivity
http://dx.doi.org/10.5772/intechopen.87738
Edited by Juan Manuel Velazquez Arcos

Contributors
George Caminha-Maciel, Nasser Edinne Benhassine, Djalil Boudjehem, Abdelnour Boukaache, Juan Manuel Velazquez Arcos, Ricardo Teodoro Páez Hernández, Alejandro Pérez Ricardez, Jaime Granados Samaniego, Alicia Cid Reborido, Anry Nersessian, Vineet Kumar, Shruti Vora, Foram Asodiya, Naveen Kumar, Anil Kumar Gangwar, Mesias Alfeus, Shiam Kannan, Eric Sakk, Ian Mackinnon, Alanoud Almutairi, Jose Alarco, Peter Brusov, Tatiana Filatova, Ashoka Reddy Komalla, Ette Harikrishna

First published in London, United Kingdom, 2021 by IntechOpen
IntechOpen is the global imprint of INTECHOPEN LIMITED, registered in England and Wales, registration number: 11086078, 5 Princes Gate Court, London, SW7 2QJ, United Kingdom
Printed in Croatia

British Library Cataloguing-in-Publication Data
A catalogue record for this book is available from the British Library

Additional hard and PDF copies can be obtained from orders@intechopen.com

Real Perspective of Fourier Transforms and Current Developments in Superconductivity
Edited by Juan Manuel Velazquez Arcos
p. cm.
Print ISBN 978-1-83962-397-4
Online ISBN 978-1-83962-398-1
eBook (PDF) ISBN 978-1-83962-399-8

For EU product safety concerns:
IN TECH d.o.o., Prolaz Marije Krucifikse Kozulić 3, 51000 Rijeka, Croatia, info@intechopen.com or visit our website at intechopen.com.

Meet the editor

Juan Manuel Velázquez Arcos obtained an MSc and Ph.D. in Physics from Universidad Nacional Autónoma de México in 1987 and 1985, respectively. From 1977 to 1986, he served as an assistant professor at the same university. Since 1986, he has been a full professor at Universidad Autónoma Metropolitana where he was also chief of the Basic Science Department from 2003 to 2007, president of the Request Judgment Committee from 2001 to 2002, member of the Academic College and of the Academic Council from 2000 to 2001, and division coordinator of the General Subjects Trunk of the Basic Sciences from 1993 to 2000. He was an invited organizer for a special session at the International Conference on Electromagnetic Advanced Applications (ICEAA) 2015 in Torino, Italy. His subjects of interest include mathematical physics, integral equations, communications theory, relativity, quantum mechanics, electromagnetic broadcasting, nanotechnology, acoustic and electromagnetic time reversal, acoustic and electromagnetic resonances, metamaterials, non-local potentials in nuclear physics, electromagnetic theory, and thermodynamics.

Contents

Preface

This book is devoted to some of the current fundamental tools on physics, communications, economy and health, that is, Fourier transforms and the treatment of superconductivity under the vision of the electromagnetic properties of both type I and type II superconductors. We begin in Chapter 1, "On the Zap Integral Operators over Fourier Transforms," with the latest applications and mathematical properties of Fourier transforms like projection operators aimed at communications and electromagnetic emissions. The authors prove that under an appropriate space generated by the solutions of the generalized inhomogeneous Fredholm equation it is possible to operate in it with integral operators like those of the quantum mechanics theory, killing and creating solutions of homogeneous or inhomogeneous Fredholm equations, which can be used to improve communications and broadcasting by selecting the type of behavior we need. Chapter 2, "Quantum Fourier Operators and Their Application" revises the structure of the Quantum Fourier Operators (QFT) and its implementation putting these concepts in their proper perspective, the authors provide a brief overview of quantum computation and provide a permutation structure for putting the QFT in the context of universal computation. In Chapter 3, "A Fast Method for Numerical Realization of Fourier Tools," the author presents a new algorithm for fast summations of truncated Fourier series. Chapter 4, "Fourier Transform Infrared Spectroscopy of the Animal Tissues," examines how animal tissues are extensively used as scaffolds for tissue engineering and regenerative therapies. They are typically subjected to a decellularization process to obtain cell-free extracellular matrix (ECM) scaffolds. It is important to identify the chemical structure of the ECM scaffolds, and Fourier transform infrared spectroscopy (FTIR) appears to be the technique of choice. The chapter presents FTIR spectra of native and decellularized buffalo aortae, buffalo diaphragms, goat skin, and native bovine cortical bone. The transmittance peaks are that of organic collagen amide A, amide B, amide I, amide II, and amide III chemical functional groups in both native and decellularized aortae, diaphragms, and skin. In bone, the transmittance peaks are that of inorganic ν_1, ν_3 PO_4^{3-}, OH^- in addition to organic collagen amide A, amide B, amide I, amide II, and amide III chemical functional groups. In Chapter 5, "Medical Image Classification Using the Discriminant Power Analysis (DPA) of Discrete Cosine Transform (DCT) Coefficients," we see the relevance of medical imaging systems in medicine. These systems assist specialists in making the final decision about a patient's condition, and strongly help in early cancer detection. The classification of mammogram images represents a very important operation to identify whether breast cancer is benign or malignant. The authors propose a new computer-aided diagnostic (CAD) system composed of three steps. In the first step, the input image is pre-processed to remove the noise and artifacts and to separate the breast profile from the pectoral muscle. This operation is a difficult task that can affect the final decision. In the second step, we propose a features extraction method based on the discrete cosine transform (DCT), where the processed images of the breast profiles are transformed by the DCT where the part containing the highest energy value is selected. Then, in the features selection step, a new most discriminative power coefficients algorithm is proposed to select the most significant features. The obtained results show the effectiveness. In Chapter 6, "Path Integral Two Dimensional Models of P– and D–Wave Superconductors and Collective

Modes," we return to superconductivity in which the main parameter that is the order parameter describes superfluids and superconductors and all their main properties.

All properties of 2D–superconductors (for example, of CuO_2 planes of HTSC) and, in particular, the collective excitations spectrum, are determined by these functionals. The authors consider all superconducting states, arising in symmetry classification of p-wave and d-wave 2D–superconductors, and calculate the full collective modes spectrum for each of these states. This will help to identify the type of pairing and the symmetry of the order parameter in HTSC and HFSC. In Chapter 7, "Periodogram Analysis under the Popper-Bayes Approach," the authors describe the Lomb-Scargle periodogram, its advantages, and pitfalls from a geometrical rather than statistical point of view. This means emphasizing more the transformation properties of the finite sampling (i.e., the available data) rather than the ensemble properties of the assumed model statistical distributions. The whole discussion is under the geophysical inverse theory point of view, the Tarantola's combination of information or the so-called Popper-Bayes approach. Finally, the authors discuss the Lomb-Scargle-Tarantola (LST) periodogram, which is an estimator of spectral content existing in irregularly sampled time series that implements these principles. Chapter 8, "Spread Option Pricing on Single-Core and Parallel Computing Architectures," introduces parallel computation for spread options using a two-dimensional Fourier transform. Spread options are multi-asset options whose payoffs depend on the difference of two underlying financial securities. Their results indicate a significant increase in computational performance when the algorithm is performed on multiple CPU cores and GPU. Moreover, the literature on spread option pricing using FFT methods documents that the pricing accuracy increases with FFT grid size while the computational speed has opposite effect. By using the multi-core/GPU implementation, the trade-off between pricing accuracy and speed is taken into account effectively. In Chapter 9, "Use of Transforms in Biomedical Signal Processing and Analysis," the authors write about low-frequency biomedical signals like electrocardiogram (ECG), photoplethysmographic (PPG), and blood pressure that need to be processed for further diagnosis and clinical monitoring. They use Fourier and wavelet transforms to reduce motion artifacts from PPG signals to produce correct blood oxygen saturation (SpO2) values. In an important contribution, FT is utilized for the generation of the reference signal for adaptive filter-based motion artifact reduction eliminating additional sensors for the acquisition of reference signals. In Chapter 10, "Insights from Systematic DFT Calculations on Superconductors," the authors present three systematic approaches to the use of Density Functional Theory (DFT) for interpretation and prediction of superconductivity in new or existing materials. Systematic calculations on conventional superconductors show that to attain a level of resolution comparable to the energy gap, two key parameters, Δk and the cut-off energy, must be optimized for a specific compound. The optimal level of resolution is achieved with k-grids smaller than the minimum reciprocal space separation between key parallel Fermi surfaces. These approaches enable estimates of superconducting properties including the transition temperature (Tc). They demonstrate these approaches for the conventional superconductors MgB2, metal substituted MgB2, and boron-doped diamond.

Juan Manuel Velázquez Arcos
Basic Sciences Department,
Metropolitan Autonomous University,
CDMX, México

Chapter 1

On the Zap Integral Operators over Fourier Transforms

Juan Manuel Velazquez Arcos,
Ricardo Teodoro Paez Hernandez, Alejandro Perez Ricardez,
Jaime Granados Samaniego and Alicia Cid Reborido

Abstract

We devote the current chapter to describe a class of integral operators with properties equivalent to a killer operator of the quantum mechanics theory acting over a determined state, literally killing the state but now operating over some kind of Fourier integral transforms that satisfies a certain Fredholm integral equation, we call this operators Zap Integral Operators (ZIO). The result of this action is to eliminate the inhomogeneous term and recover a homogeneous integral equation. We show that thanks to this class of operators we can explain the presence of two extremely different solutions of the same Generalized Inhomogeneous Fredholm equation. So we can regard the Generalized Inhomogeneous Fredholm Equation as a Super-Equation with two kinds of solutions, the resonant and the conventional but coexisting simultaneously. Also, we remember the generalized projection operators and we show they are the precursors of the ZIO. We present simultaneous academic examples for both kinds of solutions.

Keywords: integral operators, generalized inhomogeneous Fredholm equations, killer operators, evanescent waves, electromagnetic resonances

1. Introduction

Recently a new question about the solutions of integral Fredholm emerges, that is the question about the type of equation each of them solve. If we follow the steps or the clue marked by the linear second order differential equations the solutions of the inhomogeneous equation do not solve de homogeneous equation. But we have shown in a recent paper that both kind of solutions of the homogeneous and also the inhomogeneous Fredholm equations satisfy a third class of integral equation we named the Generalized Inhomogeneous Fredholm Equation (GIFE) which is only a bit different for the traditional inhomogeneous [1–3]. Even more, we can transform his appearance in a continuous form from homogeneous to inhomogeneous, but preserving his very extraordinary property: the two kinds of solutions are simultaneous solutions. This situation is quite different from differential equations but not the connection between eigenfunctions and solutions of inhomogeneous equations through the Green function [4–7]. And if we want to explain this behavior we find a founder: an integral operator which is hidden in the structure of the GIFE. There is no surprise in the fact that the new operator treats in different manner both kinds of solutions. Indeed, it seems to be natural that the new operators include the Green function and are close to the Fredholm operator [2, 3]. Before we define the ZIO operators we must underline the fact that in a broadcasting situation [8–10] we must take into accounts not only one kind of traveling waves but all the known ones because the complete description of the phenomena comes from the GIFE. Another important goal of this paper is to give an explanation of the simultaneous validity of two sets of boundary conditions that are very apart one to the other and the fact that there is a connection with other projection operators, the generalized projection operators (GPO) [11] that separates the constituents of a signal in orthogonal parts.

2. Remembering the GIFE

We remember that if we take the inhomogeneous vector integral Fredholm Eq. (1):

$$u^m(\mathbf{r},\omega) = u^{m(\circ)}(\mathbf{r},\omega) + \lambda(\omega)\int_0^\infty \mathbf{K}_n^{m(\circ)}(\omega;\mathbf{r},\mathbf{r}')u^n(\mathbf{r}',\omega)dr' \tag{1}$$

Where the kernel $\mathbf{K}_n^{m(\circ)}(\omega;\mathbf{r},\mathbf{r}')$, is the product of the interaction $A_t^m(\omega;\mathbf{r},\mathbf{s})$ (may be a non-local potential) with the free Green function $\mathbf{G}_n^{t(\circ)}(\omega;\mathbf{r},\mathbf{s})$.

And we make the ansatz of two successive approximations (a second order approach) [9], by the consideration that $\lambda(\omega)$ is a number with a very small absolute value ($|\lambda(\omega)| \ll 1$), we arrive to the integral equation we named the GIFE:

$$s^m(\mathbf{r},\omega) = s^{m(\circ)}(\mathbf{r},\omega) + \Theta^m(\mathbf{r},\omega) + \nu(\omega)\int_0^\infty \mathbf{K}_n^{m(\circ)}(\omega;\mathbf{r},\mathbf{r}')s^n(\mathbf{r}',\omega)dr' \tag{2}$$

This last equation is the one that have the property of represent a complete panorama in a broadcasting problem, that is describes both the resonant and the conventional behavior of the electromagnetic field [12].

As we have commented, Eq. (2) carries a mechanism that allows simultaneously consider both types of solution. The so called generalized source is indeed a blend of integral operators as we will see with properties we want to visualize. But first we must present the Generalized Homogeneous Fredholm Equation (GHFE) [1]:

$$y_e^m(\mathbf{r};\omega) = \eta_e(\omega)\int_0^\infty \mathbf{K}_n^{m(\circ)}(\omega;\mathbf{r},\mathbf{r}')y_e^n dr' \tag{3}$$

Eq. (3) has a special index e that mean a specific resonance [1, 4, 5, 8, 9]. Among the three Eqs. (1), (2), and (3) there are a common ingredient, for each equation we have used different names: λ, ν and η [1–3] but any of them can be incorporated to the kernel or used as an independent function or even an eigenvalue. In order to connect the homogeneous and inhomogeneous equation we must define some functions as we will see in the next sections.

3. Connection between the eigenvalues η_e and the function λ

We know that because of the Hilbert-Schmidt theory [2, 3] and more recently by our previous results [1], the solutions of Eq. (3) that is all the $y_e^m(\mathbf{r};\omega)$, form a set of orthogonal functions and then a set of eigenvalues $\eta_e(\omega)$. Thus we can relate the functions appearing in Eqs. (1) and (3) as follows:

By means of the spectral representation of Green function, [2, 3] we have:

$$\mathbf{G}_m^{n(\circ)}(\omega;\mathbf{r},\mathbf{s}) = \sum_e C_e \frac{y_e^m(\mathbf{r})y_e^n(\mathbf{s})}{\lambda - \eta_e} \tag{4}$$

And also

$$u^m(\mathbf{r},\omega) = u^{m(\circ)}(\mathbf{r},\omega) - \sum_{e=1}^{\infty}\int_0^{\infty} \frac{y_e^m(\mathbf{r},\omega)y_e^n(\mathbf{r}',\omega)}{\lambda(\omega) - \eta_e(\omega)} u^{m(\circ)}(\mathbf{r}';\omega)dr' \tag{5}$$

The orthogonality relation is

$$\int_0^{\infty} y_e^m(\mathbf{r};\omega)A^{mn}y_i^n(\mathbf{r};\omega)dr = 0 \text{ if } i \neq e \tag{6}$$

4. Conditions imposed over the homogeneous Fredholm equations

In accordance with the theory of homogeneous Fredholm integral equations [1, 2, 13], the first Fredholm minor is a two point function, like a Green function, which must comply with an integral equation:

$$\mathcal{M}^m(\mathbf{r},\mathbf{r}_0;\omega) = \eta(\omega)\Delta(\eta,\omega) + \eta(\omega)\int_0^{\infty} \mathbf{K}_n^{m(\circ)}(\omega;\mathbf{r},\mathbf{s})\mathcal{M}^n(\mathbf{s},\mathbf{r}_0;\omega)ds \tag{7}$$

Two other conditions must be satisfied:

The first is that Fredholm determinant is zero

$$\Delta(\eta,\omega) = 0 \tag{8}$$

The second that the Fredholm eigenvalue equals to one:

$$\eta(\omega) = 1 \tag{9}$$

But thanks to our second order approximation Eq. (2) we can show that other interesting conditions are satisfied, for example if we define some particular functions (and operators):

$$\Psi(\mathbf{r};\omega) \equiv \mathcal{M}^m(\mathbf{r},\mathbf{r}_0;\omega) - \Delta(\eta,\omega)u^m(\mathbf{r};\omega) \tag{10}$$

And also

$$\Psi^{(\circ)}(\mathbf{r};\omega) \equiv \Delta(\eta,\omega)\left[\eta(\omega) - u^{m(\circ)}(\mathbf{r};\omega)\right] + \Delta(\eta,\omega)[\eta(\omega) - \nu(\omega)]\int_0^\infty \mathbf{K}_n^{m(\circ)}(\omega;\mathbf{r},\mathbf{r}')u^n(\mathbf{r}';\omega)dr' \tag{11}$$

We can see that the first Fredholm minor must satisfy through Ψ the inhomogeneous equation

$$\Psi(\mathbf{r};\omega) \equiv \Psi^{(\circ)}(\mathbf{r};\omega) + \eta(\omega)\int_0^\infty \mathbf{K}_n^{m(\circ)}(\omega;\mathbf{r},\mathbf{t})\Psi(\mathbf{t};\omega)dt \tag{12}$$

In order to write Eq. (2) in terms of the solutions of Eq. (1), we can define the operator:

$$\Theta^m(\mathbf{r};\omega) \equiv \nu^2(\omega)\varepsilon(\omega)\int_0^\infty \mathbf{K}_n^{m(\circ)}(\omega;\mathbf{r},\mathbf{r}')\int_0^\infty \mathbf{K}_l^{n(\circ)}(\omega;\mathbf{r}',\mathbf{r}'')\mathrm{P}^m(\mathbf{r}'';\omega)dr''dr' \tag{13}$$

In Eq. (13) the function $\mathrm{P}^m(\mathbf{r};\omega)$ is an arbitrary negative exponential regulator. Near a resonance the two small parameters $\nu(\omega)$ and $\varepsilon(\omega)$ makes $\Theta^m(\mathbf{r};\omega)$ lesser than a second order term, so can be neglected. Far of a resonance this later function sketches the behavior of the simultaneous existence of the resonant and non-resonant solutions because in terms of $\Theta^m(\mathbf{r};\omega)$ the conventional waves satisfy the inhomogeneous equation:

$$u^m(\mathbf{r};\omega) = u^{m(\circ)}(\mathbf{r};\omega) + \Theta^m(\mathbf{r};\omega) + \nu(\omega)\int_0^\infty \mathbf{K}_n^{m(\circ)}(\omega;\mathbf{r},\mathbf{r}')u^n(\mathbf{r}';\omega)dr' \tag{14}$$

5. Defining a new class of integral operators

As we said in Section **1**, hidden in the structure of Eq. (2) there are some integral operators which allow the simultaneous existence of solutions with extremely different boundary conditions. So, let us define the Zap operators by the rules:

$$Zu^m(\mathbf{r};\omega) \equiv \mathrm{Z}\left[\mathbf{r};\omega;u^{m(\circ)}(\mathbf{r};\omega)\right]u^m(\mathbf{r};\omega) = + \Delta(\eta)\left[\eta - u^{m(0)}(\mathbf{r};\omega)\right] + \eta\int_0^\infty \mathbf{K}_n^{m(\circ)}(\omega;\mathbf{r},\mathbf{r}')u^n(\mathbf{r}';\omega)dr' \tag{15}$$

That is, the Zap operator is associated to the integral Fredholm equation satisfied by the affected solution ($u^m(\mathbf{r};\omega)$ or $y_e^m(\mathbf{r};\omega)$), from which takes the source term and the free kernel.

The same operator (15) acting over a homogeneous equation looks like

$$
\begin{aligned}
\mathrm{Z}y_e^m(\mathbf{r};\omega) \equiv \mathrm{Z}[\mathbf{r};\omega;0]y_e^m(\mathbf{r};\omega) = \\
+\Delta(\eta)[\eta - 0] \\
+\eta \int_0^\infty \mathbf{K}_n^{m(\circ)}(\omega;\mathbf{r},\mathbf{r}')y_e^n(\mathbf{r}';\omega)dr'
\end{aligned}
\tag{16}
$$

That is

$$
\mathrm{Z}y_e^m(\mathbf{r};\omega) \equiv \mathrm{Z}[\mathbf{r};\omega;0]y_e^m(\mathbf{r};\omega) = \lambda y_e^m(\mathbf{r};\omega) \tag{17}
$$

As we can see the effect of the Zap operator is to kill or eliminate the inhomogeneous term when applied to a resonant state. But this seems very artificial because we are giving indeed two parts for the complete rule. However we can build projection operators that can make the work we need.

Now, we define Zap projection operators in the next section.

6. The zap projection operators and their properties

On this section, we define the so named Zap projection operators (ZPO) which enable us to project a complex broadcasting system over a reduced resonant simplest one. The Zap operators acts over Fourier transforms [14, 15] related to integral operators.

Then, based on (15) and (17), we define de following operator:

$$
\mathbb{Z}_P\left(\mathbf{r};\omega;u^{m(\circ)}(\mathbf{r};\omega)\right) = \lim_{\eta\to 1} \mathrm{Z}\left(\mathbf{r};\omega;u^{m(\circ)}(\mathbf{r};\omega)\right) \tag{18}
$$

In order to get a display of the properties of this operator we propose a specific set of discrete antennas in the next example:

Suppose that we have p punctual sources that can be represented in the inhomogeneous term of the Fredholm equation like:

$$
u^{m(\circ)}(\mathbf{r};\omega) = \sum_{i=1}^{p} \alpha_i^n \delta(\mathbf{r} - \mathbf{r}_i)\mathbf{K}_n^{m(\circ)}(\omega;\mathbf{r},\mathbf{r}_i) \tag{19}
$$

Then, by applying the projection operator to Eq. (19) we have (remember that when $\eta = 1$ thus $\Delta = 0$):

$$
\begin{aligned}
\mathbb{Z}_P\left[\mathbf{r};\omega;u^{m(\circ)}(\mathbf{r};\omega)\right]u^m(\mathbf{r};\omega) = \lim_{\eta\to 1}\Big[\Delta(\eta)\left(\eta - \sum_{i=1}^{p} \alpha_i^n \delta(\mathbf{r}-\mathbf{r}_i)\mathbf{K}_n^{m(\circ)}(\omega;\mathbf{r},\mathbf{r}_i)\right) \\
+\eta \int_0^\infty \mathbf{K}_n^{m(\circ)}(\omega;\mathbf{r},\mathbf{r}')u^n(\mathbf{r}';\omega)dr'\Big]
\end{aligned}
\tag{20}
$$

Now, because $\eta = 1$ implies $\Delta = 0$, and because also $\nu = \lambda = 1$

$$
\mathbb{Z}_P\left[\mathbf{r};\omega;u^{m(\circ)}(\mathbf{r};\omega)\right]u^m(\mathbf{r};\omega) = y_e^m(\mathbf{r};\omega) \tag{21}
$$

In the last step we have used the fact that the solution of the remaining homogeneous equation is denoted by $y_e^m(\mathbf{r};\omega)$.

Eq. (21) says that if we take a blend of regular and resonant solutions we have:

$$\mathbb{Z}_P\left[\mathbf{r};\omega;u^{m(\circ)}(\mathbf{r};\omega)\right]\left(u^m(\mathbf{r};\omega)+y_e^m(\mathbf{r};\omega)\right)=2y_e^m(\mathbf{r};\omega) \tag{22}$$

So taken into account from Eqs. (18) until (22), we see that we have projected the original problem into a resonant one.

In analogy with $\mathbb{Z}_P$ we can define a projector over their complement:

Let us define the complementary Zap projection operator as

$$\begin{aligned}\mathbb{Z}_Q\left(\mathbf{r};\omega;u^{m(\circ)}(\mathbf{r};\omega)\right)u^m(\mathbf{r};\omega) &\equiv \lim_{\eta\to1} \mathrm{Z}^C\left(\mathbf{r};\omega;u^{m(\circ)}(\mathbf{r};\omega)\right)\\ &\equiv \lim_{\eta\to1}\{(\Delta(\eta)\eta+\nu)\int_0^\infty \mathbf{K}_n^{m(\circ)}(\omega;\mathbf{r},\mathbf{r}')u^n(\mathbf{r}';\omega)dr'\\ &+\eta u^{m(\circ)}(\mathbf{r};\omega)\}=u^m(\mathbf{r};\omega)\end{aligned} \tag{23}$$

Even we apply $\mathbb{Z}_Q$ to a resonant state:

$$\begin{aligned}\mathbb{Z}_Q\left(\mathbf{r};\omega;u^{m(\circ)}(\mathbf{r};\omega)\right)y_e^m(\mathbf{r};\omega) &\equiv \lim_{\eta\to1} \mathrm{Z}^C\left(\mathbf{r};\omega;u^{m(\circ)}(\mathbf{r};\omega)\right)\\ &= \lim_{\eta\to1}\{(\Delta(\eta)\eta+\nu)\int_0^\infty \mathbf{K}_n^{m(\circ)}(\omega;\mathbf{r},\mathbf{r}')y_e^n(\mathbf{r}';\omega)dr'\\ &+\eta u^{m(\circ)}(\mathbf{r};\omega)\}=u^m(\mathbf{r};\omega)\end{aligned} \tag{24}$$

This is because the name of the solution of the remaining inhomogeneous equation is precisely $u^m(\mathbf{r};\omega)$.

7. An academic example for conventional traveling waves

In order to convince us of the utility of the $\mathbb{Z}_P$ and $\mathbb{Z}_Q$ operators we remember that in all of our developments the kernel always is $\mathbf{K}_n^{m(\circ)}$ that only contains the free Green function $\mathbf{G}_n^{t(\circ)}(\omega;\mathbf{r},\mathbf{s})$. But then, there is no difference between the kernels of the integral equations when are referred to conventional traveling waves or to evanescent or resonant waves. This last statement allows describing in an algebraic mode the application of the Zap projection operators. In this manner we can fix our kernel in accordance with a previous example that we have presented in some place as the matrix (27).

For the case of only two source points and omitting the three components of the field lifting only one, this matrix can be for example:

But first remember that

$$\int_V A(\mathbf{r}')G_\omega^{(\circ)}(\mathbf{r};\mathbf{r}')u(\mathbf{r}';\omega)dV' \equiv \mathbf{K}^{(\circ)}(\omega)u(\mathbf{r};\omega) \tag{25}$$

In Eq. (25) $A(\mathbf{r})$ is the interaction that in the general case may contain a non-local potential, but not in our example.

$$\mathbf{K}^{(\circ)}(\omega)u(\mathbf{r};\omega) \equiv \begin{pmatrix} \mathbf{K}_{11}^{(\circ)} & \mathbf{K}_{12}^{(\circ)} \\ \mathbf{K}_{21}^{(\circ)} & \mathbf{K}_{22}^{(\circ)} \end{pmatrix} \begin{pmatrix} u_1(\mathbf{r};\omega) \\ u_2(\mathbf{r};\omega) \end{pmatrix} \equiv \mathbf{K}^{(\circ)}(\omega) \begin{pmatrix} g_1(\omega) \\ g_2(\omega) \end{pmatrix} u(\mathbf{r}) \tag{26}$$

In Eq. (26) $u(\mathbf{r})$ is a scalar function.
And then, the kernel may be

$$\mathbf{K}^{(\circ)}(\omega) = \begin{pmatrix} \dfrac{\sin(\omega-\omega_p)d}{(\omega-\omega_p)d} & -i\dfrac{\cos(\omega-\omega_p)d}{(\omega-\omega_p)d} \\ i\dfrac{\cos(\omega-\omega_p)d}{(\omega-\omega_p)d} & \dfrac{\sin(\omega-\omega_p)d}{(\omega-\omega_p)d} \end{pmatrix} \tag{27}$$

So Eq. (19) takes the form:

$$u^{(\circ)}(\mathbf{r};\omega) = \sum_{i=1}^{2} \alpha_i \delta(\mathbf{r}-\mathbf{r}_i)\mathbf{K}^{(\circ)}(\omega)\mathbf{e}_i \tag{28}$$

Where

$$\mathbf{e}_1 = \begin{pmatrix} 1 \\ 0 \end{pmatrix} \text{and} \mathbf{e}_2 = \begin{pmatrix} 0 \\ 1 \end{pmatrix} \tag{29}$$

That is

$$u^{(\circ)}(\mathbf{r};\omega) = \alpha_1\delta(\mathbf{r}-\mathbf{r}_1)\mathbf{K}^{(\circ)}(\omega)\mathbf{e}_1 + \alpha_2\delta(\mathbf{r}-\mathbf{r}_2)\mathbf{K}^{(\circ)}(\omega)\mathbf{e}_2 \tag{30}$$

The conventional waves satisfy the scalar form of Eq. (1)

$$\begin{aligned} u(\mathbf{r},\omega) &= u^{(\circ)}(\mathbf{r},\omega) \\ &+ \lambda(\omega)\int_0^\infty \mathbf{K}^{(\circ)}(\omega;\mathbf{r},\mathbf{r}')u(\mathbf{r}',\omega)dr' \end{aligned} \tag{31}$$

Or in accordance with Eq. (25)

$$\begin{aligned} u(\mathbf{r},\omega) &= u^{(\circ)}(\mathbf{r},\omega) \\ &+ \lambda(\omega)\mathbf{K}^{(\circ)}(\omega)u(\mathbf{r},\omega) \end{aligned} \tag{32}$$

Where the form of $u(\mathbf{r},\omega)$ is unknown but possibly be sketched as

$$u(\mathbf{r},\omega) = \begin{pmatrix} \dfrac{\sin(\omega-\omega_p)d}{(\omega-\omega_p)d} \\ \dfrac{\sin(\omega-\omega_p+\beta)d}{(\omega-\omega_p+\beta)d} \end{pmatrix} u(\mathbf{r}) \tag{33}$$

Now we can apply the projection operator $\mathbb{Z}_P\left[\mathbf{r};\omega;u^{(\circ)}(\mathbf{r};\omega)\right]$ to Eq. (32) and obtain

$$\mathbb{Z}_P\left[\mathbf{r};\omega;u^{(\circ)}(\mathbf{r};\omega)\right]u(\mathbf{r};\omega)=\lim_{\eta\to 1}\left[\Delta(\eta)\left(\eta-\sum_{i=1}^{2}\alpha_i\delta(\mathbf{r}-\mathbf{r}_i)\mathbf{K}^{(\circ)}(\omega)\mathbf{e}_i\right)\right.$$
$$\left.+\eta\mathbf{K}^{(\circ)}(\omega)u(\mathbf{r};\omega)\right] \tag{34}$$

Then, by putting $\eta = 1$ and $\Delta = 0$ finally

$$\mathbb{Z}_P\left[\mathbf{r};\omega;u^{(\circ)}(\mathbf{r};\omega)\right]u(\mathbf{r};\omega)=y_e(\mathbf{r};\omega) \tag{35}$$

So it is irrelevant the part of the problem concerning the two sources, it is only a problem about resonances. Our problem is now to find the resonant frequencies by taking $\mathbf{K}^{(\circ)}(\omega)$ and impose the conditions $\eta = 1$ and $\Delta = 0$.

But, what is the real advantage of the $\mathbb{Z}_P$ and $\mathbb{Z}_Q$ operators?, the answer is that the Zap operator formalism may be viewed as a test for distinguish between an expression that cannot be transformed or yes, in whatever sense between the homogeneous and inhomogeneous equations under the rules established above; if not, we can ensure that some kind of irregular things are present. In case of the positive transformation, we have the confidence that both kinds of solutions can coexists, and then we can separate the solutions for convenience as if it was a problem of two steps: homogeneous and inhomogeneous.

Now the last condition over the Fredholm determinant is

$$\Delta\begin{pmatrix}\dfrac{\sin(\omega-\omega_p)d}{(\omega-\omega_p)d}-\eta & -i\dfrac{\cos(\omega-\omega_p)d}{(\omega-\omega_p)d}\\ i\dfrac{\cos(\omega-\omega_p)d}{(\omega-\omega_p)d} & \dfrac{\sin(\omega-\omega_p)d}{(\omega-\omega_p)d}-\eta\end{pmatrix}=0 \tag{36}$$

The parameters d, η (the Fredholm eigenvalue) and ω_p (the plasma frequency) can take in principle, arbitrary values but for a specific media can be take numeric values. Now, we remember that we must also impose $\eta = 1$.

Then, Eq. (36) has two resonances:

$$\omega_1=\frac{\pi}{4d}+\omega_p \tag{37}$$

And

$$\omega_1=\frac{3\pi}{4d}+\omega_p \tag{38}$$

8. Forerunners of the zap projection operators

In Section **6** we defined a new class of integral operators we named Zap projection operators that literally cleans from a broadcasting problem the inhomogeneous part and leaves a projected homogeneous version. These operators act directly over an inhomogeneous Fredholm equation and are related to the Fredholm operators. But recently, we have defined another set of operators we called generalized projection operators (GPO) which projects a complete broadcasting signal (maybe described by a GIFE) not only into several independent mutually orthogonal signals but also can reverse the time direction as we wish. These GPO may be considered as the precursors of the Zap operators and we will see why. We remember their form:

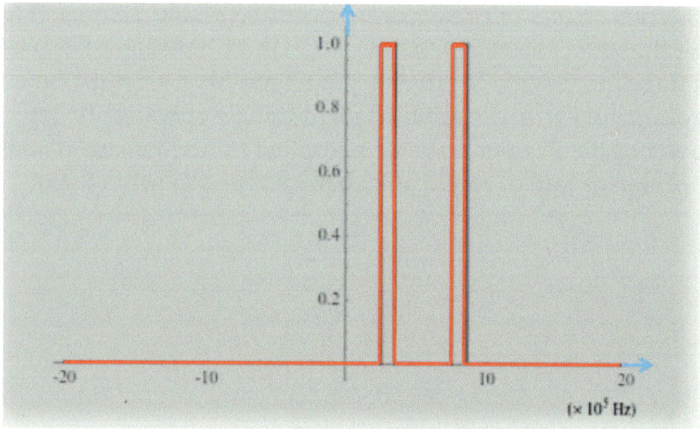

Figure 1.
The two rectangular functions $p_{4\pi\omega_e}(2(\omega-\omega_e)\omega_e)$ *and* $p_{4\pi\omega_u}(2(\omega-\omega_u)\omega_u)$.

$$\Omega^{(+)}_{e,T} S_a(t-T) = e^{i\omega_e t}\mathrm{P}^e_{S_a}(t) \quad (39)$$

And

$$\Omega^{(-)}_{e,T} S_a(t) = e^{i\omega_e(T-t)}\mathrm{P}^e_{S_a}(T-t) \quad (40)$$

In Eqs. (39) and (40) $\mathrm{P}^e_{S_a}(T-t)$ are simple projection operators [11].
Denoting the Fourier transform like

$$\mathcal{F}[f(t)] \equiv \mathrm{F}(\omega) \quad (41)$$

Then, the Fourier transform of the GPO is

$$\mathcal{F}\left[\mathrm{P}^e_{S_a}(t)e^{i\omega_e t}\right] = \sum_{n=-\infty}^{\infty} \mathrm{C}_{n,e}|2\omega_e|p_{4\pi\omega_e}(2(\omega-\omega_e)\omega_e)e^{-i(\omega-\omega_e)\pi n} \equiv F_{S_a}(\omega-\omega_e) \quad (42)$$

Where $p_{4\pi\omega_e}(2(\omega-\omega_e)\omega_e)$ is a rectangular function.
And for the convolution we have

$$\mathcal{F}\left[\mathrm{P}^e_{S_a}(t)e^{i\omega_e t} * \mathrm{P}^e_{S_b}(t)e^{i\omega_u t}\right] = F_{S_a}(\omega-\omega_e)F_{S_b}(\omega-\omega_u) \quad (43)$$

Then we see that the set of Fourier transforms of the GPOs behaves like a set of orthogonal basis functions for the frequency domain, that is, the resonant functions $y^m_e(\mathbf{r};\omega)$ as we can verify in **Figure 1**. So the GPO can be considered as the fore-runners of the Zap projection operators.

9. Conclusions

We can conclude that the Zap projection operators (ZPO) can be used as an alternative approach to the generalized projection operators (GPO) that is like an alternative for clean the evanescent signals [10] from disturbances generated by the sources and at the same time to clean the source signals from resonant solutions. We can also use the two classes of projectors in a consecutively manner. The former vision suppose that the evanescent waves [10] can be considered as part of the conventional traveling waves like an everything and that we must take away the effect of the resonances with the application of the $\mathbb{Z}_Q$ operator. In any case we have shown the power of the Fourier transform applied to mathematical analysis in broadcasting problems and to physically characterize and solve them.

Author details

Juan Manuel Velazquez Arcos*, Ricardo Teodoro Paez Hernandez,
Alejandro Perez Ricardez, Jaime Granados Samaniego and Alicia Cid Reborido
Metropolitan Autonomous University, Mexico City, Mexico

*Address all correspondence to: jmva@correo.azc.uam.mx

References

[1] J. M. Velázquez-Arcos, *Fredholm's equations for subwavelength focusing*. J. Math. Phys. Vol. 53, No. 10, 103520 (2012), doi: 10.1063/1.4759502.

[2] Harry Hochstadt, *Integral Equations*, Wiley Classics Library, ISBN-10 : 1580531741; ISBN-13 : 978–1580531740 .

[3] Witold Pogorzelsky and Ian Snnedon, *Integral Equations and their Applications*, Pergamon, ISBN-10 : 0486685225; ISBN-13 : 978–0486685229 .

[4] A. Mondragón, E. Hernández and J. M. Velázquez Arcos *Resonances and Gamow States in Non-Local Potentials*, Annalen der Physik *Volume 48, Issue 8*, 1991, PP 503–616 .

[5] R. de la Madrid, *The decay widths, the decay constants, and the branching fractions of a resonant state*, Nuclear Physics A 940 (2015) 297–310.

[6] J. M. Velázquez-Arcos, C. A. Vargas, J. L. Fernández-Chapou, A. L. Salas-Brito, *On computing the trace of the kernel of the homogeneous Fredholm's equation*. J. Math. Phys. Vol. 49, 103508 (2008), doi: 10.1063/1.3003062.

[7] R. de la Madrid, *The rigged Hilbert space approach to the Gamow states*, J. Math. Phys. Vol. **53**, No. 10, 102113 (2012), doi: 10.1063/1.4758925.

[8] J. M. Velázquez-Arcos and J. Granados-Samaniego, *Wave propagation under confinement break*, IOSR Journal of Electronics and Communication Engineering (IOSR-JECE)*e-ISSN: 2278–2834, p- ISSN: 2278–8735. Volume 11, Issue 2, Ver. I (Mar-Apr. 2016), PP 42-48www.iosrjournals.org.*

[9] J. M. Velázquez-Arcos, J. Granados-Samaniego, C.A. Vargas, *The confinement of electromagnetic waves and Fredholm's alternative*, Electromagnetics in Advanced Applications (ICEAA), 2013 International Conference pp.411–414, 9–13 Sept. 2013 doi: 10.1109/ ICEAA.2013.6632268.

[10] Xiang-kun Kong, Shao-bin Liu, Hai-feng Zhang, Bo-rui Bian, Hai-ming Li et al., *Evanescent wave decomposition in a novel resonator comprising unmagnetized and magnetized plasma layers*, Physics of Plasmas, Vol. 20, 043515 (2013); doi: 10.1063/1.4802807.

[11] J. M. Velázquez-Arcos, J. Granados-Samaniego, A. Cid-Reborido and C. A. Vargas, *The electromagnetic Resonant Vector and the Generalized Projection Operator,* IEEE Xplore, 2018 Progress In Electromagnetics Research Symposium (PIERS-Toyama), Japan, pp 1225–1232.

[12] F. Hernández-Bautista, C. A. Vargas and J. M. Velázquez-Arcos, *Negative refractive index in split ring resonators*, Rev. Mex. Fis. Vol. **59**, no. 1, pp. 139–144, January–February 2013, ISSN: 0035-00IX.

[13] von der Heydt von N. *Die Schrödinger-Gleichung mit nichtlokalem Potential. I.) Die Resolvente*, Annalen der Physik *Volume 29, Issue 4, 1973, PP 309–324.*

[14] Richard R. Goldberg, *Fourier Transforms, Cambridge University Press*. ISBN-13: 978–0521095556; ISBN-10: 0521095557.

[15] David Brandwood, *Fourier Transforms in Radar and Signal Processing*, Artech House Radar Library ISBN-10: 1580531741; ISBN-13: 978–1580531740.

Chapter 2

Quantum Fourier Operators and Their Application

Eric Sakk

Abstract

The application of the quantum Fourier transform (QFT) within the field of quantum computation has been manifold. Shor's algorithm, phase estimation and computing discrete logarithms are but a few classic examples of its use. These initial blueprints for quantum algorithms have sparked a cascade of tantalizing solutions to problems considered to be intractable on a classical computer. Therefore, two main threads of research have unfolded. First, novel applications and algorithms involving the QFT are continually being developed. Second, improvements in the algorithmic complexity of the QFT are also a sought after commodity. In this work, we review the structure of the QFT and its implementation. In order to put these concepts in their proper perspective, we provide a brief overview of quantum computation. Finally, we provide a permutation structure for putting the QFT within the context of universal computation.

Keywords: quantum Fourier transform, quantum computation, quantum circuit, entanglement, unitary operators, permutation operators

1. Introduction

The quantum Fourier transform (QFT) has been applied in a number of different contexts within the field of quantum computation [1–3]. As this operator is central to so many quantum algorithms, a major thrust of current research is directed toward its efficient implementation [4–9]. The QFT calculation is, to a degree, based upon the discrete Fourier transform (DFT) where, given a discrete sequence

$$x = \{x_0, x_2, \cdots, x_{N-1}\} \quad (1)$$

of length N, the DFT of x can be computed as

$$DFT\{x\} = Fx \quad (2)$$

with DFT matrix elements

$$F_{jk} = \frac{1}{\sqrt{N}} e^{i\frac{2\pi}{N}jk} \quad j,k = 0,1,\cdots,N-1 \quad (3)$$

Since the DFT matrix is $N \times N$, the computational complexity of computing $DFT\{x\}$ is $\mathcal{O}(N^2)$. If the input sequence length of the input sequence x can be written as $N = 2^n$ (i.e. a power of two for some positive integer, n), there exist fast

Fourier transform (FFT) implementations that can compute $DFT\{x\}$ with $\mathcal{O}(N \log N)$ complexity. While there are other FFT implementations that do not require $N = 2^n$, the 'radix-2' implementation will be the starting point as it is relevant when introducing quantum computational bases. Before elevating the DFT to its quantum description, in Section 2 we will take a brief tour of quantum computation in order to provide some necessary context. We will then, in Section 3, develop the QFT operator and discuss its quantum implementation. Finally, in Section 4, we will discuss the QFT in the context of universal computation and its formulation in terms of permutation matrices.

2. Quantum computation

A starting point for quantum computation begins with choosing a qubit representation for the computational basis [3]

$$|0\rangle \equiv \begin{pmatrix} 1 \\ 0 \end{pmatrix}, |1\rangle \equiv \begin{pmatrix} 0 \\ 1 \end{pmatrix} \tag{4}$$

This qubit basis forms a complete orthonormal set so that any single qubit quantum mechanical state can be written as the linear superposition

$$|\psi\rangle = \alpha|0\rangle + \beta|1\rangle. \tag{5}$$

where the coefficients α and β are complex scalars. If $\langle\psi|$ represents the Hermitian conjugate of $|\psi\rangle$, according to quantum mechanics, the inner product

$$\langle\psi|\psi\rangle = |\alpha|^2 + |\beta|^2 = 1 \tag{6}$$

is normalized so that ψ represents a probability density function. This implies that, at any given instance in its time evolution, a quantum system can simultaneously be in the logical states $|0\rangle$ and $|1\rangle$ with their associated probabilities $|\alpha|^2$ and $|\beta|^2$. This is in stark contrast to classical digital computation whose operations must always exclusively evaluate to a value of either 0 or 1. Quantum computation allows an algorithm to simultaneously visit *both* logical states $|0\rangle$ and $|1\rangle$ of a single qubit. If n qubits (i.e. multiple qubits) are applied, then a quantum system, in principal, has the potential to simultaneously visit 2^n logical states (again, with their associated probabilities). This exponential computational capacity is the source of quantum parallelism. However, there is a catch. Only when some observable is measured can we ascertain the current logical state of the system. Hence, quantum computers require large samples of measurements in order to build up the statistics necessary to determine the outcome of any given algorithm.

2.1 Unitary operators

The time evolution operator U associated with a quantum system must be unitary meaning that

$$U^\dagger U = I \tag{7}$$

where $U^\dagger$ is the conjugate transpose of U. A major implication of this requirement is that the forward time system evolution must (at least mathematically) be reversible. This requirement, in turn, constrains computations that are implemented by quantum operators to be ***reversible***. Therefore, logical operations such as AND, OR, and XOR (exclusive-or) would not be a quantum mechanical possibility unless some additional input information were to be preserved. This is because, in the absence of information about the input, measuring the output of these operations is not enough to ascertain the values of the inputs. Hence, these boolean processes, by themselves, are not reversible. However, there is a theory of reversible computation that can augment these logical operations so that input information is recoverable. Furthermore, much thought has gone into phrasing reversible computation in the context of unitary operators. Given the discussion so far, it is appropriate to give a short list of standard single qubit operators:

$$I = \begin{bmatrix} 1 & 0 \\ 0 & 1 \end{bmatrix}, H = \frac{1}{\sqrt{2}}\begin{bmatrix} 1 & 1 \\ 1 & -1 \end{bmatrix}, R_\phi = \begin{bmatrix} 1 & 0 \\ 0 & e^{i\phi} \end{bmatrix} \tag{8}$$

$$X = \begin{bmatrix} 0 & 1 \\ 1 & 0 \end{bmatrix}, Y = \begin{bmatrix} 0 & -i \\ i & 0 \end{bmatrix}, Z = \begin{bmatrix} 1 & 0 \\ 0 & -1 \end{bmatrix} \tag{9}$$

The reader can check that these are all unitary. As a simple example of how to apply such operators, consider the action of X on the basis vector $|0\rangle$

$$X|0\rangle = \begin{bmatrix} 0 & 1 \\ 1 & 0 \end{bmatrix}|0\rangle = \begin{bmatrix} 0 & 1 \\ 1 & 0 \end{bmatrix}\begin{bmatrix} 1 \\ 0 \end{bmatrix} = \begin{bmatrix} 0 \\ 1 \end{bmatrix} = |1\rangle \tag{10}$$

where $|0\rangle$ and $|1\rangle$ are 'swapped', indicating a form of logical inversion. H is a Hadamard transform (i.e. a DFT for a sequence of length N=2). X, Y and Z are Pauli matrices. R_ϕ is a generalization of $Z = R_\pi$ and $I = R_0$. While these are single qubit operators, the next sections discuss how they can be extended to the multiple qubit case. Amazingly, this set of quantum operators can be applied to devise some very powerful quantum algorithms (e.g. QFT computation) [3, 10].

2.2 Tensor product (Kronecker product)

The Kronecker product of an $m \times n$ matrix A with a $p \times q$ matrix B is defined to be

$$A \otimes B = \begin{bmatrix} a_{11}B & a_{12}B & \cdots & a_{1n}B \\ a_{21}B & a_{22}B & \cdots & a_{2n}B \\ \vdots & \vdots & \vdots & \vdots \\ a_{m1}B & a_{m2}B & \cdots & a_{mn}B \end{bmatrix}. \tag{11}$$

Furthermore, assuming the dimensions are compatible for matrix multiplication, the following identity often proves useful

$$(A \otimes B)(C \otimes D) = (AC) \otimes (BD) \tag{12}$$

for matrices A, B, C, D.

The computational basis can be extended to any number of qubits using the tensor product. For example, if two qubits are required for the computational space, using Eq. (2), the basis becomes

$$\begin{aligned}
|0\rangle &\equiv |00\rangle = |0\rangle \otimes |0\rangle = \begin{pmatrix} 1 \\ 0 \end{pmatrix} \otimes \begin{pmatrix} 1 \\ 0 \end{pmatrix} = \begin{pmatrix} 1 \\ 0 \\ 0 \\ 0 \end{pmatrix} \\
|1\rangle &\equiv |01\rangle = |0\rangle \otimes |1\rangle = \begin{pmatrix} 1 \\ 0 \end{pmatrix} \otimes \begin{pmatrix} 0 \\ 1 \end{pmatrix} = \begin{pmatrix} 0 \\ 1 \\ 0 \\ 0 \end{pmatrix} \\
|2\rangle &\equiv |10\rangle = |1\rangle \otimes |0\rangle = \begin{pmatrix} 0 \\ 1 \end{pmatrix} \otimes \begin{pmatrix} 1 \\ 0 \end{pmatrix} = \begin{pmatrix} 0 \\ 0 \\ 1 \\ 0 \end{pmatrix} \\
|3\rangle &\equiv |11\rangle = |1\rangle \otimes |1\rangle = \begin{pmatrix} 0 \\ 1 \end{pmatrix} \otimes \begin{pmatrix} 0 \\ 1 \end{pmatrix} = \begin{pmatrix} 0 \\ 0 \\ 0 \\ 1 \end{pmatrix}
\end{aligned} \tag{13}$$

To generalize this example for n qubits, the set of computational basis vectors can, for the sake of brevity, be labeled in base 10 as

$$\{|0\rangle, |1\rangle, |2\rangle, \cdots, |2^n - 1\rangle\}. \tag{14}$$

On the other hand, in order to highlight the qubit values, this basis can equivalently be expressed in base 2 as

$$|k_1 k_2 \cdots k_n\rangle = |k_1\rangle \otimes |k_2\rangle \otimes |\cdots \otimes |k_n\rangle \tag{15}$$

where $k_i \in \{0, 1\}$ for $i = 1, \cdots, n$. In other words, $\{k_1, k_2, \cdots, k_n\}$ represents the binary expansion

$$k = k_1 2^{n-1} + k_2 2^{n-2} + \cdots + k_{n-1} 2^1 + k_n 2^0 = \sum_{t=1}^{n} k_t 2^{n-t} \tag{16}$$

for the kth basis vector $|k\rangle \equiv |k_1 k_2 \cdots k_n\rangle$. We have chosen this bit index ordering as it will prove convenient for the QFT formulation in the next section. An equally acceptable (and, quite typical) bit index convention for an n qubit system could, for example, be $|q\rangle \equiv |q_{n-1} q_{n-2} \cdots q_1 q_0\rangle$.

Eq. (15) tells us that the n qubit basis is derived from the tensor product of single qubits. This is important to keep in mind in order to avoid confusion when using the symbol $|0\rangle$. For example, when using $n = 1$ qubit, $|0\rangle$ in decimal is equivalent to $|0\rangle$ in binary; however,, when using $n = 3$ qubits, $|0\rangle$ in decimal is equivalent to $|000\rangle$ in binary. Hence, the number of qubits n is the anchor for the relationship between Eq. (14) and Eq. (15). Assuming n qubits, there are 2^n basis vectors that can be used to construct a quantum state. Hence, all 2^n basis vectors will simultaneously evolve with their associated probabilities; again, this is the source of quantum parallelism.

2.3 Quantum circuits

One particularly useful application of Eq. (12) arises when building up n qubit quantum circuits (i.e. schematic depictions of quantum operations on qubits). For instance, assume a two qubit system $|q_1q_0\rangle$ where two unitary operators H and Z act on single qubits as

$$H|q_1\rangle, Z|q_0\rangle \tag{17}$$

and the result is desired to be combined as

$$H|q_1\rangle \otimes Z|q_0\rangle. \tag{18}$$

Eq. (12) tells us that this action is equivalent to

$$(H \otimes Z)(|q_1\rangle \otimes |q_0\rangle). \tag{19}$$

However, by construction, $|q_1\rangle \otimes |q_0\rangle = |q_1q_0\rangle$. Therefore,

$$H|q_1\rangle \otimes Z|q_0\rangle = (H \otimes Z)|q_1q_0\rangle \tag{20}$$

making it straightforward to develop multiple qubit quantum systems from unitary operators. The schematic representation of $(H \otimes Z)|q_1q_0\rangle$ is show in **Figure 1**.

With the groundwork laid for multiple qubits, it becomes possible to introduce more unitary operators that facilitate reversible computation. For example, the controlled NOT (CNOT) function can be phrased as a two qubit reversible XOR operator

$$CNOT = \begin{bmatrix} 1 & 0 & 0 & 0 \\ 0 & 1 & 0 & 0 \\ 0 & 0 & 0 & 1 \\ 0 & 0 & 1 & 0 \end{bmatrix}. \tag{21}$$

where c represents the control bit, t represents the target XOR function and $|q_1q_0\rangle = |ct\rangle$. This operator is a permutation matrix that is consistent with **Table 1** in that it swaps the $|11\rangle$ and $|10\rangle$ qubits. The XOR operation, by itself, can act as an irreversible controlled NOT operation. For the sake of quantum computation, the CNOT operator is unitary and a reversible XOR function is achieved because the control bit $|q_1\rangle$ is preserved from input to output.

There exist powerful tools for the simulation of quantum operations (referred to as '*quantum gates*') and for the rendering of multiple qubit quantum circuits [11].

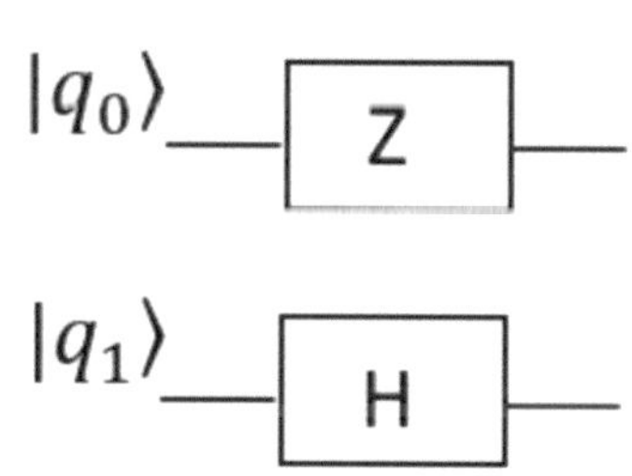

Figure 1.
Two qubit quantum circuit for $(H \otimes Z)|q_1q_0\rangle$.

c_{in}	t_{in}	c_{out}	t_{out}
0	0	0	0
0	1	0	1
1	0	1	1
1	1	1	0

Table 1.
Controlled NOT.

Figure 2 shows a schematic representation of the CNOT circuit corresponding to **Table 1**. In this circuit, the control bit is used to swap the target $|q_0\rangle$ (using an X gate) if $|q_1\rangle = |1\rangle$.

For the sake of this work, we point out that an equally valid interpretation of the quantum CNOT function can be realized if the roles of the control and target are interchanged where $|q_1q_0\rangle = |tc\rangle$ (see **Table 2**). In this case the CNOT operator becomes

$$CNOT = \begin{bmatrix} 1 & 0 & 0 & 0 \\ 0 & 0 & 0 & 1 \\ 0 & 0 & 1 & 0 \\ 0 & 1 & 0 & 0 \end{bmatrix}. \tag{22}$$

which is a permutation matrix that swaps the $|11\rangle$ and $|01\rangle$ qubits and corresponds to the circuit in **Figure 3**.

We shall have more to say about this implementation in the following sections. For now, with this brief overview of quantum computation, we can now introduce the quantum Fourier transform.

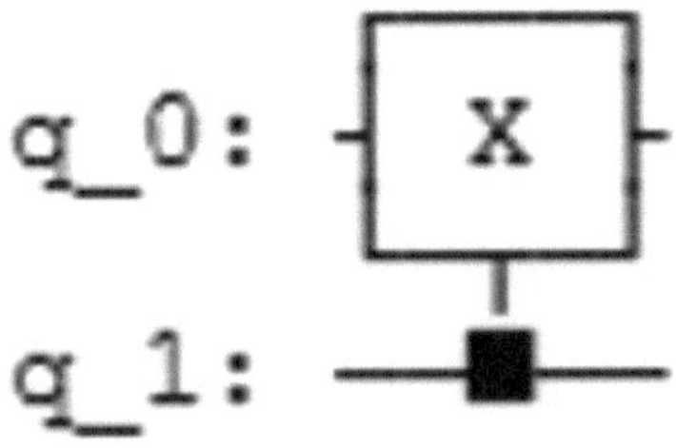

Figure 2.
Two qubit CNOT quantum circuit swap of $|11\rangle$ *and* $|10\rangle$ *using Qiskit [11].*

t_{in}	c_{in}	t_{out}	c_{out}
0	0	0	0
0	1	1	1
1	0	1	0
1	1	0	1

Table 2.
Controlled NOT where $|q_1q_0\rangle = |tc\rangle$.

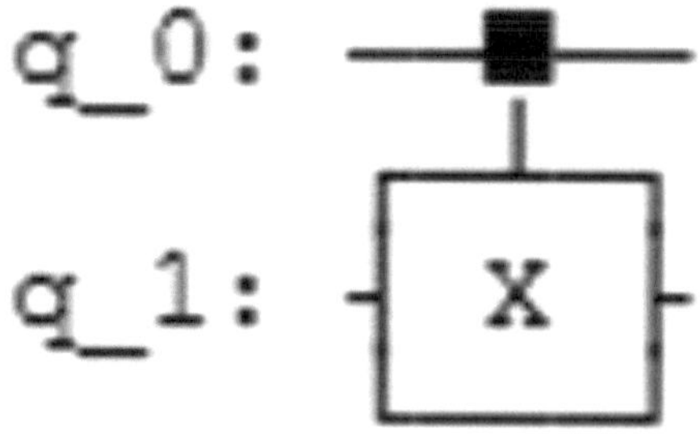

Figure 3.
Two qubit CNOT quantum circuit swap of $|11\rangle$ *and* $|01\rangle$.

3. The quantum Fourier transform

It should be clear that the DFT matrix in Eq. (2) is unitary where

$$F^{\dagger}F = I \tag{23}$$

and $F^{\dagger}$ is the Hermitian conjugate of F. Because of this unitarity, the potential for using the DFT within the context of quantum computation naturally follows. However, such an application requires a decomposition involving tensor products of unitary operations typically applied in quantum computation. As with the FFT, the choice of the decomposition dictates the algorithmic complexity. There is much introductory literature available regarding the QFT [3, 12–14]. Given a specific quantum algorithm where the QFT is applied, current research endeavors reside in attempts to improve the computational complexity [4, 7, 9, 15, 16].

The QFT matrix is defined as

$$Q = \frac{1}{\sqrt{N}} \sum_{j=0}^{N-1} \sum_{k=0}^{N-1} e^{i\frac{2\pi}{N}jk} |k\rangle\langle j|. \tag{24}$$

For example, with $N = 2^n$ and n = 1, we recover the Hadamard matrix

$$Q = \frac{1}{\sqrt{2}} \begin{bmatrix} 1 & 1 \\ 1 & -1 \end{bmatrix}, \tag{25}$$

or, for n = 2,

$$Q = \frac{1}{2} \begin{bmatrix} 1 & 1 & 1 & 1 \\ 1 & i & -1 & -i \\ 1 & 1 & 1 & -1 \\ 1 & -i & -1 & i \end{bmatrix}. \tag{26}$$

As expected, this operator is unitary where, with

$$Q^{\dagger} = \frac{1}{\sqrt{N}} \sum_{j=0}^{N-1} \sum_{k=0}^{N-1} e^{-i\frac{2\pi}{N}jk} |j\rangle\langle k|, \tag{27}$$

it should be clear that

$$\begin{aligned}
QQ^{\dagger} &= \left(\frac{1}{\sqrt{N}}\sum_{j=0}^{N-1}\sum_{k=0}^{N-1} e^{i\frac{2\pi}{N}jk}|k\rangle\langle j|\right)\left(\frac{1}{\sqrt{N}}\sum_{j'=0}^{N-1}\sum_{k'=0}^{N-1} e^{-i\frac{2\pi}{N}j'k'}|j'\rangle\langle k'|\right)\\
&= \frac{1}{N}\sum_{k'=0}^{N-1}\sum_{k=0}^{N-1}\sum_{j'=0}^{N-1}\sum_{j=0}^{N-1} e^{i\frac{2\pi}{N}\left(jk-j'k'\right)}|k\rangle\langle j|j'\rangle\langle k'|\\
&= \frac{1}{N}\sum_{k'=0}^{N-1}\sum_{k=0}^{N-1}\sum_{j'=0}^{N-1}\sum_{j=0}^{N-1} e^{i\frac{2\pi}{N}\left(jk-j'k'\right)}\delta_{j'j}|k\rangle\langle k'|\\
&= \frac{1}{N}\sum_{k'=0}^{N-1}\sum_{k=0}^{N-1}\sum_{j=0}^{N-1} e^{i\frac{2\pi}{N}j\left(k-k'\right)}|k\rangle\langle k'|\\
&= \frac{1}{N}\sum_{k'=0}^{N-1}\sum_{k=0}^{N-1}\left(N\delta_{k'k}\right)|k\rangle\langle k'|\\
&= \sum_{k=0}^{N-1}|k\rangle\langle k|\\
&= I.
\end{aligned} \tag{28}$$

In general, given a state vector

$$|\psi\rangle = \sum_{j=0}^{N-1} a_j|j\rangle \tag{29}$$

the QFT operates on $|\psi\rangle$ to form

$$\begin{aligned}
|\Psi\rangle = QFT\{|\psi\rangle\} &= QFT\left\{\sum_{j=0}^{N-1} a_j|j\rangle\right\}\\
&= \sum_{j=0}^{N-1} a_j QFT\{|j\rangle\}\\
&= \sum_{j=0}^{N-1} a_j Q|j\rangle.
\end{aligned} \tag{30}$$

Given this result, let us consider the QFT of a single n qubit basis vector $|j\rangle$ where $N = 2^n$. First, observe that while

$$\begin{aligned}
QFT\{|j\rangle\} = Q|j\rangle &= \frac{1}{2^{n/2}}\sum_{j'=0}^{N-1}\sum_{k=0}^{N-1} e^{i\frac{2\pi}{N}j'k}|k\rangle\langle j'|j\rangle\\
&= \frac{1}{2^{n/2}}\sum_{j'=0}^{N-1}\sum_{k=0}^{N-1} e^{i\frac{2\pi}{N}j'k}|k\rangle\delta_{j'j}\\
&= \frac{1}{2^{n/2}}\sum_{k=0}^{2^n-1} e^{i\frac{2\pi}{2^n}jk}|k\rangle,
\end{aligned} \tag{31}$$

given Eqs. (15) and (16), it will be more helpful to express this relation as

$$\begin{aligned}
QFT\{|j\rangle\} &= \frac{1}{2^{n/2}}\sum_{k_1=0}^{1}\sum_{k_2=0}^{1}\cdots\sum_{k_n=0}^{1} e^{i2\pi j\left(\sum_{t=1}^{n} k_t 2^{-t}\right)} |k_1k_2\cdots k_n\rangle. \\
&= \frac{1}{2^{n/2}}\sum_{k_1=0}^{1}\sum_{k_2=0}^{1}\cdots\sum_{k_n=0}^{1} e^{i2\pi j\left(k_1 2^{-1}+k_2 2^{-2}+\cdots+k_{n-1}2^{-(n-1)}+k_n 2^{-n}\right)} |k_1k_2\cdots k_n\rangle. \\
&= \frac{1}{2^{n/2}}\sum_{k_1=0}^{1}\sum_{k_2=0}^{1}\cdots\sum_{k_n=0}^{1} e^{i2\pi j\left(k_1 2^{-1}+k_2 2^{-2}+\cdots+k_{n-1}2^{-(n-1)}+k_n 2^{-n}\right)} |k_1\rangle\otimes|k_2\rangle\otimes|\cdots\otimes|k_n\rangle. \\
&= \frac{1}{2^{n/2}}\bigotimes_{v=1}^{n}\sum_{k_v=0}^{1} e^{i2\pi j(k_v 2^{-v})}|k_v\rangle.
\end{aligned} \tag{32}$$

This leads to the result that

$$Q|j\rangle = \frac{1}{2^{n/2}}\bigotimes_{v=1}^{n}\left(|0\rangle + e^{i2\pi j 2^{-v}}|1\rangle\right). \tag{33}$$

3.1 QFT qubit representation

To forge a path toward efficient implementation, it is important to recognize how Eq. (33) can be decomposed into a set of operators relevant to quantum computation (see Section 2.1). First, consider the $n = 1$ single qubit case,

$$Q|j\rangle = \frac{1}{\sqrt{2}}\left(|0\rangle + e^{i\frac{2\pi j}{2}}|1\rangle\right). \tag{34}$$

Then, for each qubit state $|j\rangle = |0\rangle, |1\rangle$, it follows that

$$\begin{aligned}
Q|0\rangle &= \frac{1}{\sqrt{2}}(|0\rangle+|1\rangle) = \frac{1}{\sqrt{2}}\begin{pmatrix}1\\1\end{pmatrix} \\
Q|1\rangle &= \frac{1}{\sqrt{2}}(|0\rangle-|1\rangle) = \frac{1}{\sqrt{2}}\begin{pmatrix}1\\-1\end{pmatrix}
\end{aligned} \tag{35}$$

as expected since $Q = H$ for the single qubit case. Hence, it should be nn surprise that the $v = 1$ contribution to Eq. (10) should be a Hadamard gate.

To handle the phase factors in the other contributions to the tensor product (where $v \geq 2$), the keen eye will recognize that the terms $e^{i2\pi j 2^{-v}}$ could lead to a unitary quantum mechanical operator. Before leveraging this observation in a QFT algorithm, it will be helpful to consider the qubit representation $|j\rangle = |j_1 j_2 \cdots j_n\rangle$. As the index v ranges from 1 to n, the index j in the term $e^{i2\pi j 2^{-v}}$ experiences successive divisions by 2 (i.e. successive right shifts of its binary representation by one bit):

$$\begin{array}{lccc}
v = 1: & j2^{-1} & \Rightarrow & j_1 j_2 \cdots j_{n-1} . j_n \\
v = 2: & j2^{-2} & \Rightarrow & j_1 j_2 \cdots j_{n-2} . j_{n-1} j_n \\
 & \vdots & & \\
v = n-1: & j2^{-(n-1)} & \Rightarrow & j_1 . j_2 \cdots j_{n-1} j_n \\
v = n: & j2^{-n} & \Rightarrow & 0.j_1 j_2 \cdots j_{n-1} j_n
\end{array} \tag{36}$$

Since these values appear in the phase factor, the integer parts will only result in integer multiples of 2π and can therefore be discarded. Eq. (33) can then be expressed as

$$QFT\{|j\rangle\} = \frac{1}{2^{n/2}}[\left(|0\rangle + e^{i2\pi j_n 2^{-1}}|1\rangle\right) \otimes \left(|0\rangle + e^{i2\pi\left(j_{n-1}2^{-1}+j_n 2^{-2}\right)}|1\rangle\right) \otimes \cdots$$
$$\cdots \otimes \left(|0\rangle + e^{i2\pi\left(j_1 2^{-1}+j_2 2^{-2}+\cdots+j_n 2^{-n}\right)}|1\rangle\right). \quad (37)$$

It is often this version of the QFT that is used as a starting point for quantum circuit implementation when $N = 2^n$ [3].

As an example, consider the two qubit case where $n = 2$ and $|j\rangle = |j_1 j_2\rangle$, then

$$\begin{aligned} Q|j\rangle &= Q|j_1 j_2\rangle \\ &= \frac{1}{2}\left(|0\rangle + e^{i2\pi j_2 2^{-1}}|1\rangle\right) \otimes \left(|0\rangle + e^{i2\pi\left(j_1 2^{-1}+j_2 2^{-2}\right)}|1\rangle\right) \end{aligned} \quad (38)$$

If we let $|j_1 j_2\rangle = |01\rangle$, then

$$\begin{aligned} Q|j\rangle &= Q|01\rangle \\ &= \frac{1}{2}\left(|0\rangle + e^{i2\pi(1)2^{-1}}|1\rangle\right) \otimes \left(|0\rangle + e^{i2\pi\left((0)2^{-1}+(1)2^{-2}\right)}|1\rangle\right) \\ &= \frac{1}{2}(|0\rangle - |1\rangle) \otimes (|0\rangle + i|1\rangle) \\ &= \frac{1}{2}(|00\rangle + i|01\rangle - |10\rangle - i|11\rangle) \end{aligned} \quad (39)$$

which corresponds to the column $|01\rangle$ entries in Eq. (26). If not already obvious, it should be emphasized that the tensor product is ***not commutative*** and that consistent qubit ordering is instrumental to the success of this calculation.

3.2 Quantum implementation

Based upon Eq. (37), it is sensible to introduce an iterable version of the R operator introduced in Section 2.1:

$$R_v = \begin{bmatrix} 1 & 0 \\ 0 & e^{\frac{i2\pi}{2^v}} \end{bmatrix}. \quad (40)$$

Furthermore, because each qubit contribution contains phase terms involving the binary expansion of j, one approach to addressing these interactions is to introduce a controlled version of R_v:

$$CR_v = \begin{bmatrix} I & 0 \\ 0 & R_v \end{bmatrix}. \quad (41)$$

This operator can be used to induce the correct phase factor as follows. Assume $tc\rangle$ is the target/control structure for single qubits $j_r j_s$ were $s > r$ in the binary representation of $|j\rangle$. Then, the following holds true

$$\begin{aligned} CR_v|j_r 0\rangle &= |j_r 0\rangle \\ CR_v|j_r 1\rangle &= e^{\frac{i2\pi}{2^v} j_r}|j_r 1\rangle \end{aligned} \quad (42)$$

Hence, the control bit determines when to introduce the phase factor involving the target bit.

The goal of this section is to introduce enough nomenclature in order to put the next section of this work in context. The reader is encouraged to visit the provided references in order to fill in the details of a generalized quantum circuit that can implement an n qubit QFT. For now, we provide an $n = 2$ qubit example to illustrate an algorithm for performing the QFT. Whatever principled series of operations is chosen, the goal of the quantum algorithm (and, hence, the associated quantum circuit) is to reproduce Eq. (11). Starting with $|j\rangle = |j_1 j_2\rangle$,

a. Apply H to $|j_1\rangle$ so that

$$\begin{aligned}|j_1\rangle \otimes |j_2\rangle &\rightarrow H|j_1\rangle \otimes |j_2\rangle \\ &= \frac{1}{\sqrt{2}}\left(|0\rangle + e^{i2\pi j_1 2^{-1}}|1\rangle\right) \otimes |j_2\rangle\end{aligned} \tag{43}$$

b. Apply CR_2 to target qubit j_1 controlled by j_2. This yields

$$\frac{1}{\sqrt{2}}\left(|0\rangle + e^{i2\pi j_1 2^{-1}}|1\rangle\right) \otimes |j_2\rangle \rightarrow \frac{1}{\sqrt{2}}\left(|0\rangle + e^{i2\pi\left(j_2 2^{-2} + j_1 2^{-1}\right)}|1\rangle\right) \otimes |j_2\rangle \tag{44}$$

c. Apply H to $|j_2\rangle$

$$\begin{aligned}\frac{1}{\sqrt{2}}\left(|0\rangle + e^{i2\pi\left(j_2 2^{-2} + j_1 2^{-1}\right)}|1\rangle\right) \otimes |j_2\rangle &\rightarrow \frac{1}{\sqrt{2}}\left(|0\rangle + e^{i2\pi\left(j_2 2^{-2} + j_1 2^{-1}\right)}|1\rangle\right) \otimes \frac{1}{\sqrt{2}}\left(|0\rangle + e^{i2\pi j_2 2^{-1}}|1\rangle\right) \\ &= \frac{1}{2}\left(|0\rangle + e^{\frac{i2\pi j}{2^2}}|1\rangle\right) \otimes \left(|0\rangle + e^{\frac{i2\pi j}{2^1}}|1\rangle\right)\end{aligned} \tag{45}$$

Comparing this result with either Eq. (33) or Eq. (37), it is clear that this algorithm, derived using quantum reversible operators, recovers the QFT from Eq. (38) with one slight difference: the bit ordering is reversed. Given n qubits, it is possible to apply $n/2$ swaps using, for example, tensor products involving an X operator (see Section 2.1) in order to reverse the bit order. Such bit reversal permutations are reminiscent of the radix-2 FFT algorithm. If one generalizes this algorithm to n qubits, it can be shown that the algorithmic complexity is $\mathcal{O}(n^2)$. With $N = 2^n$, this is a considerable improvement over $N \log N = n2^n$ for the radix-2 FFT. However, algorithmic improvements and variations have been developed that can further reduce QFT complexity to $\mathcal{O}(n \log n)$ [9, 15].

4. QFT permutations

Universal computation, by its very nature, must involve some set of permutation operators [17–20]. As with other universal gates applied in quantum computation, in this section, we show that the QFT can generate operators that have the properties of a permutation. Consider a successive application of the QFT such as $Q^2 = QQ$ and let us analyze the matrix elements of such an operation:

$$
\begin{aligned}
[QQ]_{j,k} &= \frac{1}{N}\sum_{m=0}^{N-1}\left(e^{i\frac{2\pi}{N}jm}|j\rangle\langle m|\right)\left(e^{i\frac{2\pi}{N}mk}|m\rangle\langle k|\right) \\
&= \frac{1}{N}\sum_{m=0}^{N-1} e^{i\frac{2\pi}{N}m(j+k)}\langle m|m\rangle|j\rangle\langle k| \\
&= \begin{cases} 0 & j+k \neq 0 \mod N \\ 1 & j+k = 0 \mod N \end{cases} \\
&\equiv \left[P_{Q^2}\right]_{j,k}.
\end{aligned}
\tag{46}
$$

For an n qubit system $|q_{n-1}\cdots q_1 q_0\rangle$, it should be clear that P_{Q^2} is a permutation operator that leaves the position of $|q_0\rangle$ unchanged and inverts the order of the remaining qubits to form $|q_1\cdots q_{n-1}q_0\rangle$. For example, the CNOT operator in Eq. (22) is equal to P_{Q^2} for $n = 2$

$$
CNOT = Q^2 = \begin{bmatrix} 1 & 0 & 0 & 0 \\ 0 & 0 & 0 & 1 \\ 0 & 0 & 1 & 0 \\ 0 & 1 & 0 & 0 \end{bmatrix} = P_{Q^2} \tag{47}
$$

having properties similar to that of a Sylvester shift matrix (i.e. a generalization of a Pauli matrix). It is sensible that a CNOT operation followed by a CNOT operation should result in the identity operation and, hence, that $P_{Q^2}P_{Q^2} = Q^4 = I$ (i.e. a double inversion recovers the original qubit sequence). These results can be generalized for any n. For example, with $n = 3$, Eq. (46) becomes

$$
P_{Q^2} = \begin{bmatrix} 1 & 0 & 0 & 0 & 0 & 0 & 0 & 0 \\ 0 & 0 & 0 & 0 & 0 & 0 & 0 & 1 \\ 0 & 0 & 0 & 0 & 0 & 0 & 1 & 0 \\ 0 & 0 & 0 & 0 & 0 & 1 & 0 & 0 \\ 0 & 0 & 0 & 0 & 1 & 0 & 0 & 0 \\ 0 & 0 & 0 & 1 & 0 & 0 & 0 & 0 \\ 0 & 0 & 1 & 0 & 0 & 0 & 0 & 0 \\ 0 & 1 & 0 & 0 & 0 & 0 & 0 & 0 \end{bmatrix} \tag{48}
$$

which, after the appropriate sequence of swaps, can be transformed into a Toffoli (CCNOT) gate. Hence, P_{Q^2} can be thought of as a generalization of swap permutation operators and the QFT can be phrased as its square root. For example, it is common to define a two qubit swap operator as

$$
S_w = \begin{bmatrix} 1 & 0 & 0 & 0 \\ 0 & 0 & 1 & 0 \\ 0 & 1 & 0 & 0 \\ 0 & 0 & 0 & 1 \end{bmatrix} \tag{49}
$$

along with its square root

$$\sqrt{S_w} = \begin{bmatrix} 1 & 0 & 0 & 0 \\ 0 & \frac{1}{2}(1+i) & \frac{1}{2}(1+i) & 0 \\ 0 & \frac{1}{2}(1+i) & \frac{1}{2}(1+i) & 0 \\ 0 & 0 & 0 & 1 \end{bmatrix}. \tag{50}$$

In a similar manner, Eq. (46) leads us to the following
Theorem 1 *Given the $N \times N$ inversion permutation matrix defined as*

$$\left[P_{Q^2}\right]_{j,k} = \begin{cases} 0 & j+k \neq 0 \mod N \\ 1 & j+k = 0 \mod N \end{cases}, \tag{51}$$

it follows that

$$Q = \sqrt{P_{Q^2}} \tag{52}$$

where Q is a QFT matrix.
In addition, given that $Q^4 = I$ we have the following
Corollary 1 Any algorithm that iteratively applies the QFT can result in only one of the following outcomes

a. $Q^k = \sqrt{P_{Q^2}}$ if $k = 1 \mod 4$.

b. $Q^k = P_{Q^2}$ if $k = 2 \mod 4$.

c. $Q^k = Q^{-1}$ if $k = 3 \mod 4$.

d. $Q^k = I$ if $k = 0 \mod 4$.

These results indicate a deeper connection between universal computation, permutations and the QFT. Furthermore, decomposing the QFT calculation into a product of permutations indicates a potential for reducing the computational complexity of QFT implementations.

5. Conclusions

In this work, we have revisited the quantum Fourier transform which is central to many algorithms applied in the field of quantum computation. As a natural extension of the discrete Fourier transform, the QFT can be implemented using efficient tensor products of quantum operators. Part of the thrust of current research deals with reducing the QFT computational complexity. With this goal in mind, we have phrased the QFT as a permutation operator. Future research will be directed toward quantum circuit implementation using QFT permutation operators within the context of universal computation.

Acknowledgements

This research is funded by a grant from the National Science Foundation NSF #1560214.

Author details

Eric Sakk
Morgan State University, Baltimore, MD, USA

*Address all correspondence to: eric.sakk@morgan.edu

References

[1] Shor, PW.: Polynomial-Time Algorithms for Prime Factorization and Discrete Logarithms on a Quantum Computer. SIAM J. Comput., 1997; 26: 1484–1509.

[2] Josza, R.: Quantum Algorithms and the Fourier Transform. Proc. R. Soc. Lond. A, 1998; 454:323–337.

[3] Nielsen, MA., Chuang, IL.: Quantum Computation and Quantum Information. Cambridge University Press. 2011.

[4] Barenco, A., Ekert, A., Suominen, KA., Torma, P. : Approximate quantum Fourier transform and decoherence. Phys. Rev. A, 1996; 54.

[5] Fowler, A., Hollenberg, LCL. : Scalability of Shor's algorithm with a limited set of rotation gate. Phys. Rev. A, 2004; 70.

[6] Pavlidis,A., Gizopoulos, D.: Fast Quantum Modular Exponentiation Architecture for Shor's Factorization Algorithm. Quantum Information and Computation, 2014; 14.

[7] Prokopenya,AN.: Approximate Quantum Fourier Transform and Quantum Algorithm for Phase Estimation. International Workshop on Computer Algebra in Scientific Computing, 2015; 391–405.

[8] Ruiz-Perez, L., Garcia-Escartin, JC.: Quantum arithmetic with the quantum Fourier transform. Quantum Inf. Process., 2017; 16.

[9] Nam, Y., Su, Y., Maslov, D.: Approximate quantum Fourier transform with O(n log(n)) T gates. NPJ Quantum Information, 2020; 6(26).

[10] Barenco,A., Bennett,CH., Cleve, R., DiVincenzo,DP., Margolus, N., Shor,P., Sleator,T., Smolin,J.A., Weinfurter, H. : Elementary gates for quantum computation. Phys. Rev. A, 1995; 52.

[11] Open-Source Quantum Development. https://qiskit.org/ [Accessed: 1 September 2020]

[12] Quantum Fourier Transform. https://qiskit.org/textbook/ch-algorithms/quantum-fourier-transform.html [Accessed: 1 September 2020]

[13] QC - Quantum Computing Series. https://medium.com/@jonathan_hui/qc-quantum-computing-series-10ddd7977abd [Accessed: 1 September 2020]

[14] Camps, D., Van Beeumen, R., Yang, C.: Quantum Fourier Transform Revisited.Numerical Linear Algebra with Applications. 2020.

[15] Hales,L.,Hallgren,S.: An Improved Quantum Fourier Transform Algorithm and Applications. Proceedings 41st Annual Symposium on Foundations of Computer Science, 12-14 Nov. 2000, Redondo Beach, CA, USA.

[16] Wang, SP., Sakk, E.: Quantum Algorithms: Overviews, Foundations, and Speedup. ICCSP 2021, Zhuhai, China; January 8-10, 2021.

[17] DiVincenzo,DP.: Two-bit gates are universal for quantum computation. Phys. Rev. A, 1995. 51:1015–1022.

[18] Planat,M., Ul Haq,R.: The Magic of Universal Quantum Computing with Permutations. Advances in Mathematical Physics, 2020.

[19] de Almeida,AAA., Dueck,GW., daSilva,ACR.: CNOT Gate Optimizations via Qubit Permutations. Journal of Low Power Electronics, 2019; 15:182–192.

[20] Ouyangab,Y.,Shen,Y.,Chen,L.: Faster quantum computation with permutations and resonant couplings. Linear Algebra and its Applications, 2020; 592:270–286.

Chapter 3

A Fast Method for Numerical Realization of Fourier Tools

Anry Nersessian

Abstract

This chapter presents new application of author's recent algorithms for fast summations of truncated Fourier series. A complete description of this method is given, and an algorithm for numerical implementation with a given accuracy for the Fourier transform is proposed.

Keywords: Fourier series, Fourier transforms, approximation, over-convergence phenomenon, numerical methods

1. Introduction

One of the classical tools of Mathematics is the apparatus of Fourier series, based on the orthogonal system $\{e^{i\pi kx}\}, k = 0, \pm 1, \pm 2, \ldots$ complete in $L_2[-1,1]$. However, in practice it is extremely limited because of the poor approximation of piecewise smooth functions. Thus, in the case when a function has discontinuity points (taking into account also discontinuities at the ends of the interval $[-1,1]$), an intense oscillation arises in their neighborhood, and the uniform convergence is absent (the Gibbs phenomenon). This leads to a slow L_2-convergence on the entire segment $[-1,1]$. The corresponding phenomenon is observed in the case of Fourier interpolation. Classical methods of summation truncated Fourier series (see, for example, [1], Chapter 3) do not actually change the situation.

Since the beginning of the last century, a huge amount of research has been carried out, devoted to the methods of effective summation of the Fourier series. Let us briefly dwell on the works to overcome this situation (for detiles see, for example, articles [2–13] and their links).

The pioneer of "overcoming the Gibbs phenomenon" is A.N.Krylov, who at the dawn of the 20th century (see [2]) proposed methods that he later developed in the monograph [3]. In particular, he proposed the following approach. Let a piecewise smooth function f is given on the segment $[-1,1]$ with Fourier coefficients $\{f_s\}, s = 0, \pm 1, \ldots, \pm n, \ n \geq 1$, and with the following jumps of the function f or its derivatives until degrees of the order $q \geq 1$ at the points $\{a_k\}, -1 < a_1 < \cdots < a_m = 1, 1 < m < \infty$,

$$
\begin{aligned}
A_{p,k}(f) = f^{(k)}(a_p - 0) - f^{(k)}(a_p + 0),& \\
k = 0, 1, \cdots, q \geq 0, p = 1, \cdots, m.&
\end{aligned}
\tag{1}
$$

In the neighborhoods of other points we assume that $f \in C^{q+1}$. Let us construct a function $g = g(x), x \in [-1,1]$ with Fourier coefficients $\{g_s\}, 0 \leq |s| \leq n$, which has the same jumps at the same points, and $g \in C^{q+1}$ at the neighborhoods of other points.

Known the jumps (1), one can construct, e.g. piecewise-polynomial g. As a result, 2-periodic extension of the function $F = (f - g)$ is q times continuously differentiable on whole axis, and then

$$f(x) - g(x) = \sum_{s=-n}^{n} (f_s - g_s) e^{i\pi s x} + r_s(x),$$

where $r_s(x) = o(s^{-q}), s \to \infty, x \in [-1, 1]$.

Therefore, taking into account only first $(2n + 1)$ Fourier coefficients and truncating the remainder term r_s, it is possible to approximate f in the form

$$f(x) \simeq f_n(x) \stackrel{def}{=} g(x) + \sum_{k=-n}^{n} (f_s - g_s) e^{i\pi s x} \quad (2)$$

However, the computing of the jumps $\{A_{sk}(f)\}$ directly by the function f sharply limits the scope of practical application of the method of A. N. Krylov.

The "spectral" method proposed by Kurt Eckhoff in [4] (1993) turned out to be more practical, as it is based only on the use of coefficients $\{f_s\}$ (see also [5]). It is easy to obtain using the integration in parts the following asymptotic representation of the Fourier coefficients:

$$f_s = -\frac{1}{2} \sum_{p=1}^{m} e^{-i\pi s a_p} \sum_{k=0}^{q-1} \frac{A_{p,k}}{(i\pi s)^{k+1}} + r_s, r_n = o(s^{-q}), s \to \infty. \quad (3)$$

As a function g from (2) K. Eckhoff used those Bernoulli polynomials $\{B_k(x)\}, x \in [-1, 1], k \geq 0$, which Fourier coefficients $\{b_{k,s}\}$ have the following simple form

$$b_{k,s} = \begin{cases} 0, & s = 0, \\ \dfrac{(-1)^{s+1}}{2(i\pi s)^{k+1}}, & s \neq 0, k = 1, 2, \ldots \end{cases} \quad (4)$$

Denoting $B_0(x) = 1$, polynomials $\{B_k(x)\}, k = 0, 1, \ldots, n, \ x \in [-1, 1]$ compose a basis on the space of polynomials of degree n. Bernoulli polynomials extended to the real axis with period 2 as piecewise-smooth functions.

According to Krylov's scheme, the sequence

$$F_n(x) \stackrel{def}{=} \sum_{p=1}^{m} \sum_{k=0}^{q} A_{p,k} B_k(x - a_p - 1) + \sum_{s=-n}^{n} \omega_s e^{i\pi s x}, \quad (5)$$

where the quantities $\{\omega_s\}$ are given explicitly, converges to f with the rate $o(n^{-q}), n \to \infty$.

K. Eckhoff suggests to find approximate values of jumps $\{\tilde{A}_{p,k} \approx A_{p,k}\}$ by solving the following system of linear equations with the Vandermonde matrix, obtained by principal part (1) choosing the indexes $s = s_k, k = 1, 2, \ldots, m(q + 1), \theta n \leq |s_k| \leq n$, $0 < \theta = const < 1$

$$f_s = -\frac{1}{2} \sum_{p=1}^{m} e^{-i\pi s a_p} \sum_{k=0}^{q} \frac{\tilde{A}_{p,k}}{(i\pi s)^{k+1}}, s = s_1, s_2, \ldots, s_{m(q+1)} \quad (6)$$

Further, the problem is solved by the Krylov method. It is natural to call this acceleration scheme the Krylov-Eckhoff method **(KE-method)**. Over the past two to three decades, in the KE-method scheme, not only polynomials have been used as functions of g.

In [14], for a smooth function f, a combination of the solution of Eq.(6) and the Pade approximation (applied to the asymptotic expansion of the Fourier coefficients) is implemented there. As a result of numerical experiments it turned out that, as a rule, such an algorithm is "almost exact" on certain infinite-dimensional spaces, although only a finite number of Fourier coefficients is used. In [15], a similar phenomenon was discovered for Bessel series, and in [16] for some eigenfunction expansions. The theoretical substantiation of this phenomenon for Fourier series and the corresponding algorithm were published in [17–19]. Our goal is to use this method in the case of one-dimensional Fourier transforms. Let us start with a complete description of the required algorithms from [17].

2. Acceleration of convergence of Fourier series

2.1 Basic definitions

The classical definition of the partial sums of the Fourier series is based on a gradual increase in the frequencies of the Fourier system. Below we use a more general notations from the work [17].

Definition 1. *Let us call the truncated Fourier series any sum of the form*

$$S_n(x) \stackrel{def}{=} \sum_{k \in D_n} f_k \exp(i\pi k x), \quad x \in [-1, 1], \tag{7}$$

where

$$f_s = \frac{1}{2}\int_{-1}^{1} f(x) e^{-i\pi s x} dx, \quad s = 0, \pm 1, \pm 2, \ldots \tag{8}$$

are Fourier coefficients of f, and $D_n = \{d_k\}, k = 1, \ldots, n,$ is a set of n different integers ($n \geq 1$). We will assume that $D_0 = \emptyset$.

Definition 2. *Let $n \geq 1$ be a fixed integer. Consider a system of functions $U_n = \{\exp(i\pi\lambda_k x)\}, \lambda_k \in \mathbb{C}, x \in [-1, 1], k = 1, 2, \ldots, n,$ where $\{\lambda_k\}$ are arbitrary parameters. Consider the linear span $Q_n = span\{U_n\}$. We call a function $q \in Q_n$ as quasi-polynomial of degree at most n.*

It is easy to see that $q \in Q_n$, if and only if when either $q(x) \equiv 0$ or $q(x) = \sum_k P_{\beta_k}(x) \exp(i\pi\lambda_k x)$, where polynomials $P_{\beta_k}(x) \not\equiv 0$ have the exact degree β_k and $m = \sum_k (1 + \beta_k) \leq n$. The number m will be considered below as the degree of the quasi-polynomial q.

Definition 3. *We call the system $\{\lambda_k\} \subset \mathbb{C}$ as parameters of the quasi-polynomial $q(x) = \sum_k P_{\beta_k}(x) \exp(i\pi\lambda_k x) \in Q_n$ and the number $1 + \beta_k$ as multiplicity of the parameter λ_k.*

Definition 4. *Let parameters $\{\lambda_k\}, k = 1, \ldots, n$ are [1] fixed. We denote by $Q_n(\{\lambda_k\})$ the set of all corresponding $q(x) \in Q_n$.*

Remark 1. *The set of quasi-polynomials $q(x) \in Q_n$ is invariant with respect to a linear change of the variable x. The set of quasi polynomials $q(x) \subset Q_n(\{\lambda_k\})$ is invariant with respect to a shift of the variable x.*

For a fixed integer $n \geq 1$, consider a set of non-integer parameters $\{\lambda_k\} \subset \mathbb{C}, k \in D_n$, and the following infinite sequence

[1] Unless otherwise stated, we will not exclude the repetition of values of λ_k in a set $\{\lambda_k\}$.

$$t_{r,s} \stackrel{def}{=} (-1)^{s-r}\left(\prod_{\substack{p\in D_n\\ p\neq r}} \frac{s-p}{r-p}\right)\prod_{k\in D_n}\frac{r-\lambda_k}{s-\lambda_k}, \quad r\in D_n, \ s=0,\pm 1, \dots. \tag{9}$$

Remark 2. *The last product in (3) is invariant with respect to the numbering of the parameters $\{\lambda_k\}$. Here this numbering is tied to the set $k\in D_n$.*

In the general case, the parameters $\{\lambda_k\}$ (i.e., and $t_{r,s}$) may depend on n. In order not to complicate the notation, we will indicate this dependence as needed.

Further we denote (see [17] for details)

$$\begin{aligned} T_r(x) &\stackrel{def}{=} \exp(i\pi r x) + \sum_{s\notin D_n} t_{r,s}\exp(i\pi s x), r\in D_n, \ x\in[-1,1], \\ f(x) \simeq F_n(x) &\stackrel{def}{=} \sum_{r\in D_n} f_r T_r(x), R_n(x) \stackrel{def}{=} f(x) - F_n(x) \end{aligned} \tag{10}$$

The system $\{T_r(x), 1/2\exp(i\pi r x)\}, r\in D_n$, is biorthogonal on the segment $x\in[-1,1]$ and L_2-error of approximation $f(x)\simeq F_n(x)$ can be found from the formula

$$\|R_n\|^2 = \sum_{s\notin D_n}\left\| f_s - \sum_{r\in D_n} f_r t_{r,s}\right\|^2 \tag{11}$$

Here we confine ourselves to the one-dimensional case. A similar universal algorithm for multivariate truncated Fourier series was proposed and numerically implemented in [1].

The following obvious formula ($x\in[-1,1]$)

$$\exp(i\pi\lambda x) = \sum_{s=-\infty}^{\infty} \operatorname{sinc}(\pi(s-\lambda))\exp(i\pi s x), \quad \lambda\in\mathbb{C}, \tag{12}$$

where $\operatorname{sinc}(z) = \sin(z)/z, \operatorname{sinc}(0) = 1, z\in\mathbb{C}$, plays a key role in the future. This Fourier series can be repeatedly differentiated by the parameter λ.

2.2 The case of integer or multiple parameters

To include the case of a quasi-polynomial $q\in Q_n$ containing some integer parameters from $\{\lambda_k\}$, consider the following quasi-polynomials associated with formula (12) and sequence (9)

$$\begin{aligned} &\Lambda_{j,k}(x) \stackrel{def}{=} \\ &\frac{i^{2-k}}{2\pi^{k-1}(k-1)!}\frac{d^{k-1}}{d\lambda_j^{k-1}}\left(\csc(\pi\lambda_j)\exp(i\pi\lambda_j x)\right) = \\ &\sum_{s=-\infty}^{\infty}\frac{(-1)^s\exp(i\pi s x)}{2\left(i\pi(s-\lambda_p)\right)^k}, \\ &\lambda_j\in\mathbb{C}, \ j\in D_n, \ k\geq 1, \ x\in[-1,1]. \end{aligned} \tag{13}$$

Lemma 1. *If λ_j is an integer parameter, then*

$$\Lambda_{j,k}(x) = B_k(x)\exp(i\pi\lambda_j x) \tag{14}$$

where $B_k(x)$ is a Bernoulli polynomial (see (4)).

Since the Bernoulli polynomial $B_k(x)$ corresponds to one k -multiple parameter $\lambda_j = 0$ in the KE - method, formula (14) can be considered as a generalization of Bernoulli polynomials to the case of any integer λ_j .

Remark 3. *Like the Bernoulli polynomials in the KE- method, quasi-polynomials* $\{\Lambda_{j,k}\}, k \geq 1$ *can be decomposed into a Fourier series with coefficients (8), for which* $f_s = O\left(|s|^{-k}\right), s \to \infty$.

If in the sequence (9) parameters $\{\lambda_k\}$ and its corresponding multiplicities $\{n_k\}$ are known, then we will use the representation $(s = 0, \pm 1, \dots)$

$$t_{r,s} = (-1)^{s-r}\left(\prod_{\substack{p \in D_n \\ p \neq r}} \frac{s-p}{r-p}\right)\prod_{j \in D_m}\left(\frac{r-\lambda_j}{s-\lambda_j}\right)^{n_j}, \quad (9^*) \tag{15}$$

where $r \in D_n, D_m \subset D_n, \{n_q\}$ are corresponding positive integers, and $\sum_{j \in D_m} n_j = n, \lambda_p \neq \lambda_q$ if $p \neq q$. However, this sequence is still defined only for non-integer numbers in $\{\lambda_p\}$.

Consider now the possibility of including in the consideration of some integer parameters in (10). First of all, note that if for some $j \notin D_n, \lambda_j$ is integer, then $t_{r,\lambda_j} = \infty$. So here the system $\{T_r\}$ from (10) does not exist.

If λ_j are integers for $j, j \in D_n$, then, firstly, it is natural to accept $t_{r,r} = 1$, and secondly, we notice that for $s \neq r$ the number of the products in $t_{r,s}$ are reduced.

Finally, we note that in the latter case, the sequence $t_{r,s}$ can be represented as a sum of simple fractions with respect to $s \in \mathbb{C}$. Bearing in mind *Lemma 1*, it is not difficult to proof, that.

Lemma 2. *Let in* (9^*) *the parameters* $\{\lambda_j\}, j \in D_m$, *be given and the subset* Λ *consists of all integer parameters from* $\{\lambda_j\}$. *Then.*

i. *If there is an* $\lambda_k \in \Lambda, \lambda_k \notin D_n$, *then the system* $\{T_r\}$ *from (10) does not exist.*

ii. *If* $\Lambda = (\Lambda_1 \cup \Lambda_2) \subset D_n$, *where* $\Lambda_1 \subset D_m$ *contains only different integers and* $\Lambda_2 \subset D_m$ *contains only multiple integers, then*

$$T_r(x) = \exp(i\pi r x), r \in \Lambda_1. \tag{16}$$

With a $\lambda_r \in \Lambda_2$ *we can apply Lemma 1 and also find the explicit form of functions* $\{T_r(x)\}$.

iii. *If* $\Lambda \neq \emptyset$ *and* $\Lambda \neq D_n$ *then the system* $\{T_r(x)\}, r \in D_m \backslash \Lambda$ *is determined based on the sequence, similar to (15), in which* $n_j = 1, \forall j$, *and n is replaced by* $(n-p)$ *where p is the number of elements* Λ.

2.3 Explicit form of the system $\{T_r\}$

The following result is a generalization of Theorem 1 in [17] to the case that in formula (15) there are integer values among the parameters $\{\lambda_k\}$.

Theorem 1. *Suppose the sequence (15) is given and the possible integer parameters in* $\{\lambda_k\}$ *satisfy the condition* $\lambda_k \in D_m$. *Then the corresponding functions* $\{T_r\}$ *are quasi-polynomials and have the following explicit form*

$$T_r(x) = \sum_{j \in D_m} \sum_{k=1}^{n_j} c_{r,j,k} \; \Lambda_{j,k}(x), r \in D_n, \;\; x \in [-1, 1] \tag{17}$$

where

$$c_{r,j,k} = \frac{(-1)^{r+1} (i\pi)^k \prod_{p \in D_m} (r - \lambda_p)^{n_p}}{2(n_j - k)! \prod_{\substack{p \in D_n \\ p \neq r}} (r - p)}$$

$$\frac{d^{n_j - k}}{d\lambda_j^{n_j - k}} \left(\frac{\prod_{\substack{p \in D_n \\ p \neq r}} (\lambda_j - p)}{\prod_{\substack{p \in D_n \\ p \neq j}} (\lambda_j - \lambda_p)^{n_p}} \right)$$

2.4 The adaptive algorithm and over-convergence phenomenon

In the main Theorem of paper [17], the phenomenon of over-convergence was theoretically substantiated. The basis of this result was the following **Algorithm** $\mathfrak{A}$.

Let $S_n(x)$ be a truncated Fourier series (see (7)). Consider formulas (9)–(10) with symbolic (not numerical) parameters $\{\lambda_q\}, q \in D_n$. Let us now choose a new set of integer numbers $\tilde{D}_n = \{\tilde{d}_k\}, k = 1, \dots, n, \tilde{D}_n \cap D_n = \emptyset$. To determine the parameters $\{\lambda_q\}, q \in D_n$, we additionally use Fourier coefficients $\{f_s\}, s \in \tilde{D}_n$, and solve the following system of equations (compare with (11))

$$f_s - \sum_{r \in D_n} f_r t_{r,s} = 0, s \in \tilde{D}_n \tag{18}$$

regarding parameters $\{\lambda_q\}$. Note that Eq. (18) is essentially nonlinear. If the solution exists, then the system $\{T_r\}$ from (4) is used to approximate f by F_n (see (10)).

Here we show how Algorithm $\mathfrak{A}$ can be step-by-step realized using given $2n$ Fourier coefficients $\{f_{d_k}\} \cup \{\{f_{\tilde{d}_k}\}\}$ (see (7)).

Step 1. Using the representation (9), we formally bring the left side of (18) to a common denominator, and fix the conditions $\lambda_q \neq \tilde{d}_k, k = 1, 2, \dots, n$. Then Eq. (18) will take the form $(s \in \tilde{D}_n)$

$$f_s \prod_{q \in D_n} (s - \lambda_q) = \sum_{r \in D_n} \frac{(-1)^{s - d_r} f_r}{\operatorname{sinc}(d_r - \lambda_r)} \left(\prod_{\substack{p \in D_n \\ p \neq d_r}} \frac{s - d_p}{d_r - d_p} \right) \prod_{q \in D_n} (d_r - \lambda_q). \tag{19}$$

Step 2. Using the Vieta's formula decompose the products in (18) containing the parameters $\{\lambda_k\}$ and denote

$$e_k = (-1)^k \sum_{i_1 < i_2 < \dots < i_k} \lambda_{i_1} \lambda_{i_2}, \dots, \lambda_{i_k}, \;\; k = 1, 2, \dots, n.$$

It is not difficult to see that, as a result, Eq.(18) will take the form of a linear system of equations with respect to the variables $\{e_k\}$.

Step 3. Solve resulting equation by the least square method.

Step 4. Again, according to the Vieta's formulae find all roots$\{z_k\}, k = 1, 2, \dots, n$ of the corresponding polynomial $P(z) = \sum_{p=0}^{n} e_p z^{n-p}, \; e_0 = 1$, and put $\{\lambda_k\} = \{z_k\}$.

Step 5. Realize the approximation$f \simeq F_n$ (see (10) and (17)).

Remark 4. *Least square method provides only one solution in **Step 3**.*

The following theorem is the main result of the paper [17].

Theorem 2. *{**The phenomenon of the over-convergence**}. Let $f \in Q_n$ (see Definition 2) and the sets D_n, $\tilde{D}_n$ and the Fourier coefficients $\{f_s\}, s \in D_n \cup \tilde{D}_n$, of the f be given. Denote by Λ the set of integer parameters in the representation $f = \sum_k P_{m_k}(x) \exp(i\pi\mu_k x)$. In order for the approximation by Algorithm $\mathfrak{A}$ to be exact (that is $f(x) \equiv F_n(x), x \in [-1, 1]$), it is necessary and sufficient that $\Lambda \subseteq D_n \cup \tilde{D}_n$.*

It is now natural to formulate the following definition of the acceleration of the convergence of a truncated Fourier series.

Definition 5. *Let $f \in L_2[-1, 1]$ and the sets D_n, $\tilde{D}_n$ and the Fourier coefficients $\{f_s\}, s \in D_n \cup \tilde{D}_n$, of the f be given. Applying Algorithm 1, we get parameters $\{\lambda_k\}, k \in D_n$. The accelerating the convergence of truncated Fourier series*

$$S_{2n}(x) = \sum_{k \in D_n \cup \tilde{D}_n} f_k \exp(i\pi k x), \quad x \in [-1, 1], \tag{20}$$

we define by formula (17).

Remark 5. *There are no Fourier coefficients $\{f_r\}, r \in \tilde{D}_m$ in formula (17). These coefficients are used in Algorithm $\mathfrak{A}$ to an optimal choice of parameters $\{\lambda_k\}, k \in D_m$.*

3. An numerical algorithm for Fourier transform

Consider the Fourier transform F of a function$f \in L_2(-\infty, \infty)$ the form

$$F(\omega) = \frac{1}{\sqrt{2\pi}} \int_{-\infty}^{\infty} f(t) e^{it\omega} dt, \quad \omega \in (-\infty, \infty) \tag{21}$$

3.1 Algorithm construction scheme

The method of the item **2** can be applied by linear change of variable to function f defined on any finite segment. For a fixed $h > 0$ formula (21) can be rewritten as

$$F(\omega) = \frac{1}{\sqrt{2\pi}} \sum_{k=-\infty}^{\infty} \int_{k-h}^{k+h} f(t) e^{it\omega} dt \tag{22}$$

On the other hand, our method allows one to approximate function$f \simeq \sum f_r T_r$ on each segment $(k - h, k + h), \forall k,$ using Algorithm $\mathfrak{A}$ and formula (17).

Our idea is as follows. First, based on the required 1 error, you can leave a fnite number of intervals. Second, on the remaining intervals, the integrals are calculated explicitly. Really, each function $T_r(x)$ is a linear combination of functions of the form $\exp(i\pi\lambda_1 x)$ or its derivative with respect to λ of a certain order. For example, if $k = 0$, then for functions $\exp(i\pi\lambda_1 x)$, we have (see (12), here $k = 0$)

$$\int_{-h}^{h} \exp(i\pi\lambda_1 t) e^{it\omega} dt = 2h \,\mathrm{sinc}(h(\pi\lambda_1 + \omega))$$

On the complex plane $s \in \mathbb{C}$, this is an entire function of exponential type. It is easy to see that this is true in the general case.

Remark 6. *Depending on function* f *in formula (21), interval* $(-\infty, \infty)$ *can be split into a finite number of unequal intervals instead of partition (22).*

Let us turn to the main details of the algorithm for such a numerical implementation of the Fourier transform for a piecewise smooth function f.

3.2 Implementation notes

To make our approach understandable to a wide range of specialists, when developing an applied algorithm, we will assume that (see formulas (7–19)) for a fixed $n \geq 1, D_n = \{0, 1, \dots, n-1\}, \tilde{D}_n = \{-n, -n+1, \dots, -1\}$.

Consider now the following when writing code to implement the one-dimensional Fourier transform.

1. The computer used must be installed with programs that can perform symbolic mathematical operations as well as numerical integration (for example, as system Wolfram Mathematica, see [19]). It is desirable to use the highest possible accuracy of calculations.

2. The singularities of piecewise-smooth function f (discontinuity points of the function and the discontinuity of the derivatives of low order) must be included in the partition of interval $(-\infty, \infty)$ (see **Remark 6**).

3. The computer errs or freezes most often due to type $0/0$ uncertainty. The point is that in practice, due to rounding errors, when the parameters coincide, they are often not fixed (see formula (15)). This must be taken into account also in the presence of integer parameters (see item 2.2).

For example, let in the process of computer operation at the output of Algorithm $\mathfrak{A}$ we have $\lambda_1 = 1.00012, \lambda_2 = 1.00031$. At this point, so that the calculations do not stop, it is necessary to make following correction in the code: $\lambda_1 = \lambda_2 = 1$. Of course, here we mean the permissible error 10^{-3} when computer precision is 10^{-5} (see above item **1**).

1. During the operation of the proposed algorithm for Fourier transform, L_2- error can be monitored every time the adaptive algorithm is applied. In practice, it has the same order as the L_2-error for the Fourier transform.

3.3 A simple example

Let us explain the details on the following example of an approximate finding of the Fourier transform (see notation (21)).

$$f(x) = e^{-4\left|x-\frac{1}{3}\right|} + \frac{i}{2} e^{-4\left(x-\frac{1}{2}\right)^2}, \quad x \in (-\infty, \infty) \tag{23}$$

$$F(\omega) = \frac{4\sqrt{\frac{2}{\pi}} e^{\frac{i\omega}{3}}}{\omega^2 + 16} + \frac{i e^{-\frac{1}{16}\omega(\omega - 8i)}}{4\sqrt{2}}. \quad \omega \in (-\infty, \infty) \tag{24}$$

Our goal is to provide a final value of a L_2 - error that does not exceed 10^{-2}. With the help of numerical integration (in this case, explicitly), one can make sure that

$$\sqrt{\int_{\frac{3}{2}}^{\infty} |f(x)|^2 dx + \int_{-\infty}^{-2} |f(x)|^2 dx} = 4. \times 10^{-3}$$

and when deleting these semi-infinite intervals, further actions should lead to an error no greater than 5.99×10^{-3}.

On the other hand, we see that function f is infinitely differentiable everywhere except the point $x = 1/3$, where its derivative has a jump. Therefore (see item **2**), we can consider the following partition into two intervals: $I_1 = (-2, 1/3), I_2 = (1/3, 3/2)$. Method of item **3.1** at $n = 3$ was applied to each of these intervals.

Monitoring showed (see item **4**) that L_2- error for I_1 is $1. \times 10^{-3}$, and for I_2 is 1.44×10^{-3}. We see that this is less than allowed above.

The total L_2- error turned out to be 4.38×10^{-3}. As for the uniform error, it is reached approximately at point $\omega = 0.185$ and is equal to 1.04×10^{-3}.

These calculations were performed using system Wolfram Mathematica 9.

4. Conclusion

Those who have mastered the technique of using Algorithm $\mathfrak{A}$ can easily obtain a similar method of sine and cosine Fourier transforms. Despite the fact that with an increase in n the complexity of the corresponding algorithms increases markedly, modern computers allow you to perform the necessary actions with increased bit depth. The proposed method for finding the one-dimensional Fourier transform with a given accuracy can be useful in such different fields of activity as mathematical physics, astronomy, engineering, economics, biology, etc.

There are good prerequisites for other classical (in a broad sense) Fourier tools for efficient applications based on the method of item **2** (see articles [14, 15]).

Author details

Anry Nersessian
Institute of Mathematics of National Academy of Sciences of Armenia, Yerevan, Republic of Armenia

*Address all correspondence to: nersesyan.anry@gmail.com

References

[1] A. Zygmund, Trigonometric Series, Vol. I, Cambridge, at the University Press, 1959

[2] A. N. Krylov, About approximate calculations. Lectures given in 1906 (in Russian), St. Petersburg, Typolithography of K. Birkenfeld, 1907

[3] A. N. Krylov, Lectures on approximate computing (in Russian), St. Petersburg, Printing house of Yu.N. Erlikh, 1911

[4] K. S. Eckhoff, Accurate and efficient reconstruction of discontinuous functions from truncated series expansions, Math. Comp. v. 61,N204, 745–763, 1993

[5] K. S. Eckhoff, Accurate reconstruction of functions of finite regularity from truncated Fourier series expansions, Math. Comp. v. 64,N210, 671–690, 1995

[6] C. Lanczos, Discourse of Fourier Series, Edinburgh, Oliver and Boyd, 1966

[7] W. B. Jones and G. Hardy, Accelerating convergence of trigonometric approximations, Math. Comp., 24, 547–560, 1970

[8] A. Nersessian and A. Poghosyan, Accelerating the Convergence of Trigonometric Series, Central European Journal of Math., vol. 4:3,435–448, 2006

[9] A. Barkhudaryan, R. Barkhudaryan and A. Poghosyan, Asymptotic Behavior of Eckhoff's Method for Fourier Series Convergence Acceleration, Analysis in Theory and Applications, vol. 23 (3), 1–15, 2007

[10] B. Adcock, Gibbs phenomenon and its removal for a class of orthogonal expansions, BIT, vol. 51(1), 7–41, 2011

[11] B. Adcock, Gibbs phenomenon and its removal for a class of orthogonal expansions, BIT, vol. 51(1), 7–41, 2011

[12] A. Poghosyan, On an auto-correction phenomenon of the Krylov-Gottlieb-Eckhoff method, IMA Journ. of Numer. Analysis, v. 31 (2), pp. 512–527, 2011

[13] A. Nersessian, Quasi-polynomials of Bernoulli type and acceleration of convergence of Fourier series (in Russian). Reports of the National Academy of Sciences of Armenia, vol.104 (5), pp.280–286, 2004

[14] A. Nersessian, Acceleration of convergence of Fourier-Bessel series for piecewise smooth functions (in Russian).

[15] A. Nersessian, Accelerated convergence of eigenfunction expansions (in Russian). Reports of the National Academy of Sciences of Armenia, vol 107 (2), pp. 124–131,2007

[16] Anry Nersessian, On an Over-Convergence Phenomenon for Fourier series. Basic Approach. Armen. J. Math., V. 10, N. 9, pp. 1–22, 2018

[17] Anry Nersessian, A correction to the article "On an Over-Convergence Phenomenon for Fourier series. Basic Approach". Armen. J. Math., V.11, N. 2 , pp. 1–2, 2019

[18] A. Nersessian, Fourier tools are much more powerful than commonly thought. Lobachevskii Journal of Mathematics, vol.40, N.8 , pp.1122–1131, 2019.

[19] S. Wolfram, The Mathematica Book, fifth edition, Wolfram Media, 2003

Chapter 4

Fourier Transform Infrared Spectroscopy of the Animal Tissues

Vineet Kumar, Shruti D. Vora, Foram A. Asodiya, Naveen Kumar and Anil K. Gangwar

Abstract

Animal tissues are extensively used as scaffolds for tissue engineering and regenerative therapies. They are typically subjected to decellularization process to obtain a cell-free extracellular matrix (ECM) scaffolds. It is important to identify chemical structure of the ECM scaffolds and Fourier transform infrared (FTIR) appears to be a technique of choice. In this chapter, FTIR spectra of native and decellularized buffalo aortae, buffalo diaphragms, goat skin, and native bovine cortical bone are presented. The transmittance peaks are that of organic collagen amide A, amide B, amide I, amide II and amide III chemical functional groups in both native and decellularized aortae, diaphragms and skin. In bone, the transmittance peaks are that of inorganic ν_1, ν_3 PO_4^{3-}, OH^- in addition to organic collagen amide A, amide B, amide I, amide II and amide III chemical functional groups. These important transmittance peaks of the tissue samples will help researchers in defining the chemical structure of these animal tissues.

Keywords: buffalo aorta, buffalo diaphragm, bovine bone, goat skin, Fourier transform infrared spectroscopy

1. Introduction

The extracellular matrix (ECM) scaffolds primarily composed of structural collagen protein are widely used in tissue engineering and regenerative medicine [1 15]. These are usually prepared from animal tissues by decellularization process. Decellularization is the process of removal of native cells from animal tissue, leaving behind a three-dimensional network of ECM proteins while preserving the bioactivity and mechanics of the tissue. In the decellularization process, animal tissues are subjected to physical, enzymatic and chemical treatments. Physical methods of decellularization include freezing, direct pressure, sonication, and agitation [16]. Enzymatic techniques of decellularization include the use of protease (trypsin) [1–5, 8, 10, 12–15], endonucleases and exonucleases. Chemical methods of decellularization include the use of acids and alkalis (acetic acid, peracetic acid, hydrochloric acid, sulfuric acid, ammonium hydroxide), nonionic detergents (Triton X-100), ionic detergents (sodium dodecyl sulfate, sodium deoxycholate, Triton X-200) [1–15], zwitterionic detergents (3-[(3-cholamidopropyl)dimethylammonio]-1-propanesulfonate, sulfobetaine-10, sulfobetaine-16), organic solvent (Tri(n-butyl)phosphate) [3, 10], hypertonic and hypotonic solutions [2, 3, 8, 10, 13, 15], and chelating agents (EDTA). These reagents at higher concentrations extensively disrupt the structural proteins of ECM scaffolds and make it impossible to analyze by routine techniques [17]. Fourier transform infrared (FTIR) spectroscopy is one of the preferred technique for identification of biomolecules through the study of their characteristic vibrational movements [11, 13, 14, 18]. This technique is simple, reproducible, nondestructive to the tissue, and only small amounts of tissue (micrograms to nanograms) with a minimum preparation are required. In addition, this technique also provides molecular-level information allowing investigation of functional groups, bonding types, and molecular conformations. The characteristic peaks in FTIR spectra are molecule specific and provide direct information about biochemical composition. This chapter highlights the application of FTIR spectroscopy for characterization of native and decellularized buffalo aortae, buffalo diaphragms, goat skin, and native bovine cortical bone.

2. Materials

2.1 Chemicals and reagents

Sodium dodecyl sulfate (SDS), Trypsin, Sodium chloride (NaCl), Phosphate-buffered saline (PBS), Ethylene diaminetetraacetic acid (EDTA), Sodium azide (NaN_3), Gentamicin, Potassium bromide (KBr) powder. All the chemicals and reagents used were of the high purity and obtained from Sigma-Aldrich (St. Louis, MO, USA) unless mentioned otherwise. All solutions were prepared fresh using deionized water and analytical grade chemicals at room temperature (unless indicated otherwise).

2.2 Equipments

Magnetic stirrer (C-MAG HS7, IKA, USA), Analytical digital lab balance (Citizen Enterprises, Delhi, India), Fourier transform infrared spectrophotometer (FTIR 8400 s Shimadzu Corporation, Tokyo, Japan), Bard Parker blade number 24, Autoclaved sterile dissecting scissors, Autoclaved sterile jar (Borosil, India), Sterile measuring cylinders (Borosil, India), Dishes (Borosil, India) and Protective equipment such as surgical gloves and surgical autoclaved instruments were used.

2.3 Tissue samples

Fresh cadaver buffalo aorta (**Figure 1A**), buffalo diaphragm (**Figure 1B**) and goat skin (**Figure 1C**) collected in chilled (4°C) sterile 1X PBS (pH 7.4) containing 0.016% gentamicin (antibiotic), 0.0205% EDTA (proteolytic inhibitor) and 0.1% NaN_3 (antimycotic) were our study materials. A cortical bone collected from the anterior diaphysis of the right femur of an adult cadaver Gir cow was also used.

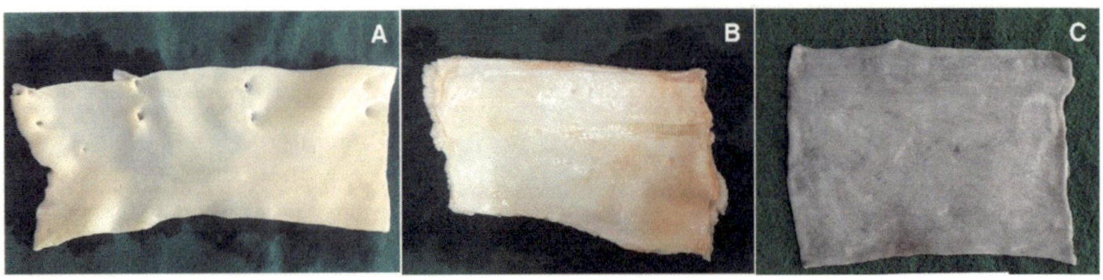

Figure 1.
Gross images of buffalo aorta (a), buffalo diaphragm (B), and goat skin (C).

3. Methods

3.1 Fourier transform infrared spectroscopy of the buffalo aorta

The *aorta,* an elastic artery, has a trilaminar structure consisting of a tunica intima, media, and adventitia. The media comprises cellular elements (including smooth muscle cells) and structural proteins (notably collagen and elastin) that form the ECM. Before being used in regenerative therapies, cellular elements of the aorta should be removed by decellularization process. The decellularization of fresh posterior aorta from deceased donor buffalo was completely achieved by treatment with 1% SDS for 24 hours followed by 0.25% trypsin for 2 hours and again by 1% SDS for 24 hours [19]. Both the native and decellularized aortae were characterized by FTIR spectroscopy [14]. Herein, one milligram of each freeze-dried tissues were mixed with pure dry KBr powder in 1:10 ratio, and pelleted. The FTIR spectra were recorded by an infrared spectrophotometer in the 500–4000 cm^{-1} wave number spectral range with a spectral resolution of 2 cm^{-1} and 45 scans. **Figure 2** illustrates the FTIR spectra of native and decellularized aortae. The transmittance peaks indicated the presence of organic collagen amide A, amide B, amide I, amide II and amide III chemical functional groups in both native and decellularized aortae. The amide A band (3294 cm^{-1}) is associated with H-bonded N-H stretching [11, 13, 20] and was found at 3282.95 cm^{-1} for native aorta and 3280 cm^{-1} for decellularized aorta. The amide B band (2953 and 2928 cm^{-1}) is related to CH_2 asymmetric stretching [11, 13, 21] and was observed at 2958.9 cm^{-1} for native aorta and 2954.08 cm^{-1} for decellularized aorta. The amide I band (1641–1658 cm^{-1}) is associated with C=O hydrogen bonded stretching [11, 13, 22] as recorded at 1658.84 cm^{-1} for native aorta and 1658.84 cm^{-1} for decellularized aorta. The amide II (1539–1546 cm^{-1}) is associated with C-N stretching and N-H in plane bending from amide linkages, including wagging vibrations of CH_2 groups from the glycine backbone and proline sidechains [11, 13, 23] in native aorta and decellularized aorta appeared at 1526.71 cm^{-1} and 1529.60 cm^{-1}, respectively. The amide III (NH bend) band was found at 1282.55 cm^{-1} for NA and 1230.69 cm^{-1} for decellularized aorta [11, 13, 24].

3.2 Fourier transform infrared spectroscopy of the buffalo diaphragm

The diaphragm is a dome shaped structure, composed of muscle surrounding a central tendon, which separates the thoracic and abdominal cavities. Before

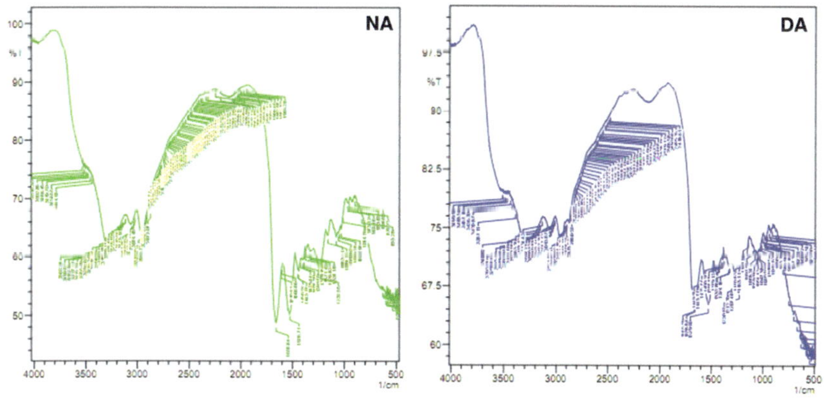

Figure 2.
FTIR spectra showing transmittance peaks of native aorta (NA) at 1282.55, 1526.71, 1658.84, 2958.9 and 3282.95 cm^{-1}; and decellularized aorta (DA) at 1230.69, 1529.60, 1658.84, 2954.08 and 3280 cm^{-1}.

the clinical application, fresh tendinous portion of diaphragm from deceased donor buffalo was decellularized using 2% SDS solution for 48 hours and FTIR spectroscopic characterization was performed [11]. Herein, one milligram of each freeze-dried native and decellularized diaphragms were mixed with pure dry KBr powder in 1:10 ratio, and pelleted. The FTIR spectra were recorded by an infrared spectrophotometer in the 500–4000 cm^{-1} wave number spectral range with a spectral resolution of 2 cm^{-1} and 45 scans. The FTIR spectra of native and decellularized diaphragms are illustrated in the **Figure 3**. The transmittance peaks indicated the presence of organic collagen amide A, amide B, amide I, amide II and amide III chemical functional groups in both native and decellularized diaphragms. The amide A band is associated with H-bonded N-H stretching [20] and was found at 3386.15 cm^{-1} for native diaphragm and 3343.71 cm^{-1} for decellularized diaphragm. The amide B band is related to CH_2 asymmetric stretching [21] and was observed at 2955.04 cm^{-1} for native diaphragm and 2954.08 cm^{-1} for decellularized diaphragm. The amide I band is associated with C=O hydrogen bonded stretching [22] as recorded at 1657.87 cm^{-1} for native diaphragm and 1649.19 cm^{-1} for decellularized diaphragm. The amide II is associated with C-N stretching and N-H in plane bending from amide linkages, including wagging vibrations of CH_2 groups from the glycine backbone and proline side-chains [23] in native diaphragm and decellularized diaphragm appeared at 1535.39 cm^{-1} and 1534.11 cm^{-1}, respectively. The amide III band was found at 1238.34 cm^{-1} for native diaphragm and 1220.02 cm^{-1} for decellularized diaphragm confirming presence of hydrogen bonds [24].

3.3 Fourier transform infrared spectroscopy of the goat skin

The goat skin consists of two layers; superficial epidermis, composed of stratified squamous keratinized epithelium and underlying dermis, composed of dense, irregular connective tissue mainly collagen fibers. Skin appendages such as hair follicles, sebaceous and sweat glands were found in the dermis. The deepithelialization of fresh goat skin was completely achieved by treatment with 0.25% trypsin in 4 mol/L NaCl for 8 hours. Further treatment with 2% SDS for 48 hours demonstrated complete decellularization of the cellular dermis [13]. Native, deepithelialized and decellularized goat skins were characterized by FTIR spectroscopy [13]. One milligram of each freeze-dried native, deepithelialized and decellularized skins were mixed with pure dry KBr powder in 1: 10 ratio, and pelleted. The FTIR spectra were recorded by an infrared spectrophotometer in the 500–4000 cm^{-1}

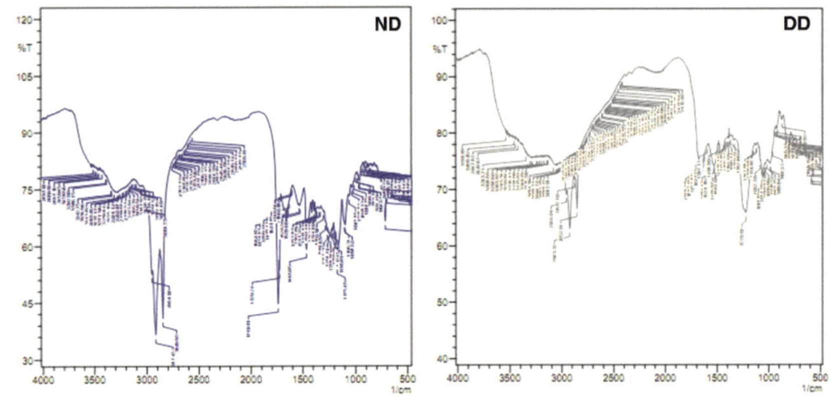

Figure 3.
FTIR spectra showing transmittance peaks of the native diaphragm (ND) at 1238.34, 1535.39, 1657.87, 2955.04 and 3386.15 cm^{-1}; decellularized diaphragm (DD) at 1220.02, 1534.11, 1649.19, 2954.08 and 3343.71 cm^{-1}.

wave-number spectral range with a spectral resolution of 2 cm^{-1} and 45 scans. The FTIR spectra of native, deepithelialized and decellularized skins are illustrated in the **Figure 4**. The transmittance peaks indicated the presence of organic collagen amide A, amide B, amide I, amide II and amide III chemical functional groups in native, deepithelialized and decellularized skins. The peaks at 3288.74 cm^{-1} for native skin, 3289.70 cm^{-1} for deepithelialized skin and 3306.10 cm^{-1} for decellularized skin corresponds to the amide A band (3294 cm^{-1}) of collagen is due to N-H stretching vibrations when the N-H group of the peptide is involved in hydrogen bonds [20]. The peaks at 2936.72 cm^{-1} for native skin, 2953.12 cm^{-1} for deepithelialized skin and 2953.12 cm^{-1} for decellularized skin corresponds to the amide B band (2953 and 2928 cm^{-1}) is due to asymmetric stretching of the CH_2 stretching vibration [21]. The peaks at 1657.87 cm^{-1} for native skin, 1658.84 cm^{-1} for deepithelialized skin and 1666.55 cm^{-1} for decellularized skin corresponds to the amide I band (1641–1658 cm^{-1}) is due to the stretching vibration of the peptide carbonyl group (-C=O) along the polypeptide backbone [22]. The peaks at 1546.96 cm^{-1} for native skin, 1530.57 cm^{-1} for deepithelialized skin and 1547.93 cm^{-1} decellularized skin corresponds to the amide II (1500–1560 cm^{-1}) which arises from the N-H bending vibration coupled to C-N stretching [23]. Amide III band was found at 1236.41 cm^{-1} for NCS, 1238.34 cm^{-1} for DCS, 1238.34 cm^{-1} for CADM, and 1238.34 cm^{-1} for BSC confirming the presence of hydrogen bonds [24].

3.4 Fourier transform infrared spectroscopy of the bovine bone

Bone is a composite biomaterial mainly composed of organic collagen fibers (chiefly type I collagen) and inorganic hydroxyapatite $[Ca_{10}(PO)_6(OH)_2]$ crystals. Both the components of bone have specific chemical signatures and, consequently, distinctive infrared spectra at the molecular level [18]. The FTIR spectroscopy of the bovine bone tissue was described in a recent study in which one milligram of powdered native cortical bone was mixed with pure dry KBr powder in 1:10 ratio, and pelleted. The FTIR spectrum was recorded by an infrared spectrophotometer

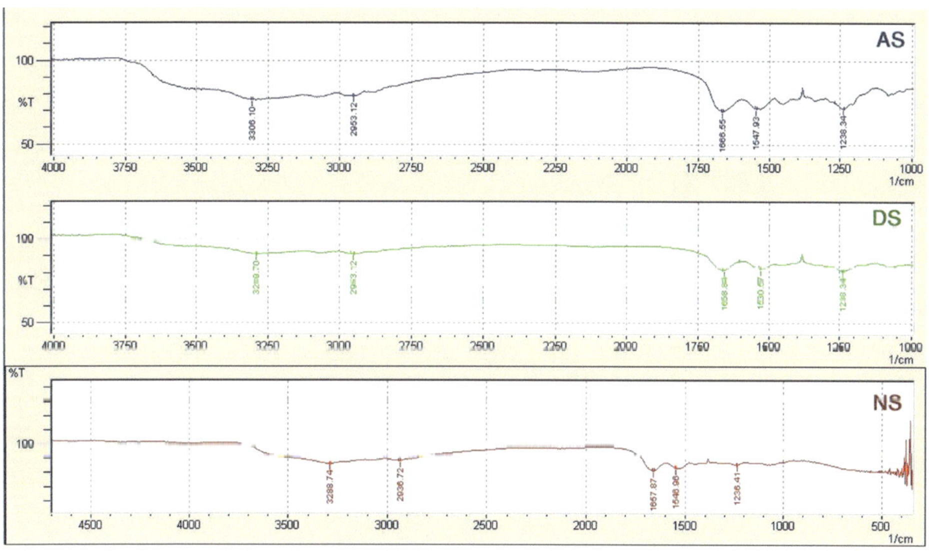

Figure 4.
FTIR spectra showing peaks of native goat skin (NS) at 1236.41, 1546.96, 1657.87, 2936.72, 3288.74 cm^{-1}, deepithelialized goat skin (DS) at 1238.34, 1530.57, 1658.84, 2953.12, 3289.70 cm^{-1} and decellularized goat skin (AS) at 1238.34, 1547.93, 1666.55, 2953.12, 3306.10 cm^{-1}.

in the 500–4000 cm^{-1} wave number spectral range with a spectral resolution of 2 cm^{-1} and 45 scans [18]. The FTIR bone spectrum can be roughly separated into two regions where the organic and inorganic components have distinct peaks (**Figure 5**). The peak at 931.65 cm^{-1} assigned to the ν_1 phosphate band [25–27] is due to the symmetric stretching vibration of the apatitic phosphate ion (PO_4^{3-}) of hydroxyapatite [27, 28]. The peak at 1032.92 cm^{-1} corresponds to the ν_3 phosphate band [29, 30] is due to the asymmetric stretching vibration of the PO_4^{3-} [27, 31]. The ν_3 phosphate peak at 1032.92 cm^{-1} is coming from the nonstoichiometric hydroxy-apatite which may contain CO_3^{2-} or HPO_4^{2-} or both in the apatite [27, 31]. With age, the concentration of CO_3^{2-} increases during apatite maturation, the amount of labile HPO_4^{2-} decreases, keeping the Ca/(C + P) atomic ratio almost constant [32]. The peaks at 1235.45 cm^{-1}, 1530.51 cm^{-1} and 1671.37 cm^{-1} correspond to the amide group, and they originate from the collagen [27]. The peak at 1235.45 cm^{-1} corresponds to the amide III results from mixed C-N stretch and N-H in-plane bend with additional contributions from C-C_α stretch [29]. The peak at 1530.51 cm^{-1} corresponds to the amide II which arises from the combined effect of C-N stretch and N-H in-plane bending [27, 31]. The peak at 1671.37 cm^{-1} corresponds to the amide I is due to the stretching vibration of the peptide carbonyl group (-C=O) along the polypeptide backbone [27]. The peak at 2879.82 cm^{-1} is asymmetric CH_2 stretch and it arises from the organic component [27]. The peak at 2965.82 cm^{-1} corresponds to the symmetric CH_2 and asymmetric CH_3 stretch of the organic component [27]. The peak at 3079.46 cm^{-1} corresponds to the amide B is due to asymmetric stretching of the CH_2 stretching vibration and the absorption due to the CH_2 alkyl chain [27]. The weak intensity peak at 3281.99 cm^{-1} corresponds to the amide A band of collagen is due to N-H stretching vibrations when the N-H group of the peptide is involved in hydrogen bonds [27]. The broad peak at 3536.60 cm^{-1} is attributed to the presence of the OH^- group [27, 33]. The transmittance peaks indicated the presence of inorganic ν_1, ν_3 PO_4^{3-}, OH^- in addition to organic collagen amide A, amide B, amide I, amide II and amide III chemical functional groups in the bovine cortical bone (**Figure 5**).

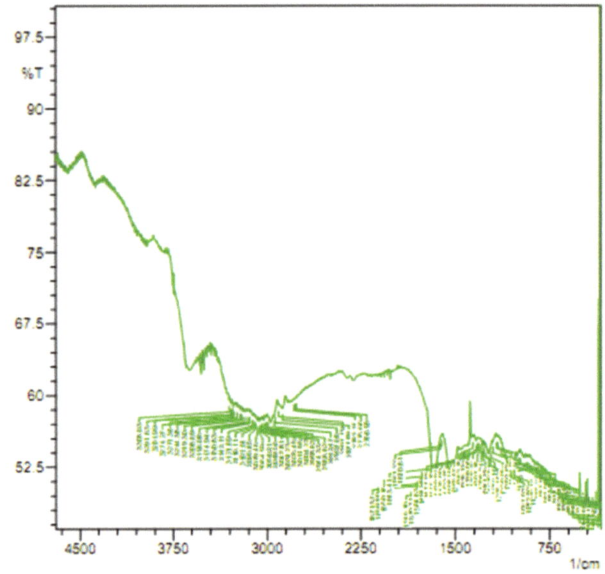

Figure 5.
FTIR spectrum of the native cortical bone showing transmittance peaks at 931.65, 1032.92, 1235.45, 1530.51, 1671.37, 2877.89, 2966.62, 3079.46, 3281.99, 3615 cm^{-1}.

4. Conclusions

The FTIR spectra of native and decellularized buffalo aortae, buffalo diaphragms, goat skin, and native bovine cortical bone are presented. The transmittance peaks are that of organic collagen amide A, amide B, amide I, amide II and amide III chemical functional groups in both native and decellularized aortae, diaphragms and skin. In bone, the transmittance peaks are that of inorganic CO_3^{2-}, ν_1, ν_3 PO_4^{3-}, OH^- in addition to organic collagen amide A, amide B, amide I, amide II and amide III chemical functional groups. These important transmittance peaks of the tissue samples will help researchers in defining the chemical structure of these animal tissues.

Author details

Vineet Kumar[1*], Shruti D. Vora[2], Foram A. Asodiya[2], Naveen Kumar[3] and Anil K. Gangwar[4]

1 Department of Veterinary Surgery and Radiology, College of Veterinary and Animal Sciences, Sardar Vallabhbhai Patel University of Agriculture and Technology, Meerut, Uttar Pradesh, India

2 Department of Veterinary Surgery and Radiology, College of Veterinary Science and Animal Husbandry, Junagadh Agricultural University, Junagadh, Gujarat, India

3 Division of Surgery, ICAR-Indian Veterinary Research Institute, Izatnagar, Uttar Pradesh, India

4 Department of Surgery, College of Veterinary Science and Animal Husbandry, Acharya Narendra Deva University of Agriculture and Technology, Ayodhya, Uttar Pradesh, India

*Address all correspondence to: bharadwaj374@gmail.com

References

[1] Kumar, V, Devarathnam, J, Gangwar, A. K, Kumar, N, Sharma, A. K, Pawde, A. M, & Singh, H. Use of acellular aortic matrix for reconstruction of abdominal hernias in buffaloes. Veterinary Record (2012)., 170, 392.

[2] Kumar, V, Kumar, N, Mathew, D. D, Gangwar, A. K, Saxena, A. C, & Remya, V. Repair of abdominal wall hernias using acellular dermal matrix in goats. Journal of Applied Animal Research (2013)., 41(1):117-120.

[3] Kumar, V, Gangwar, A. K, Mathew, D. D, Ahmad, R. A, Saxena, A. C, & Kumar, N. Acellular dermal matrix for surgical repair of ventral hernia in horses. Journal of Equine Veterinary Science (2013)., 33, 238-243.

[4] Kumar, V, Kumar, N, Gangwar, A. K, & Saxena, A. C. Using acellular aortic matrix to repair umbilical hernias of calves. Australian Veterinary Journal (2013)., 91, 251-253.

[5] Kumar, V, Kumar, N, Singh, H, Mathew, D. D, Singh, K, & Ahmad, R. A. An acellular aortic matrix of buffalo origin crosslinked with 1-ethyl-3-3-dimethylaminopropylcarbodiimide hydrochloride for the repair of inguinal hernia in horses. Equine Veterinary Education (2013)., 25, 398-402.

[6] Kumar, V, Kumar, N, Gangwar, A. K, & Singh H. Comparison of acellular small intestinal matrix (ASIM) and 1-ethyl-3-(3-dimethylaminopropyl) carbodiimide crosslinked ASIM (ASIM-EDC) for repair of full-thickness skin wounds in rabbits. Wound Medicine (2014)., 7, 24-33.

[7] Kumar, V, Kumar, N, Gangwar, A. K, Singh, H, & Singh, R. Comparative histological and immunological evaluation of 1,4-butanediol diglycidyl ether crosslinked versus noncrosslinked acellular swim bladder matrix for healing of full-thickness skin wounds in rabbits. Journal Surgical Research (2015)., 197, 436-446.

[8] Gangwar, A. K, Kumar, N, Devi, K. S, Kumar, V, & Singh R. Primary chicken embryo fibroblasts seeded 3-D acellular dermal matrix (3-D ADM) improve regeneration of full thickness skin wounds in rats. Tissue & Cell (2015)., 47, 311-322.

[9] Kumar, V, Gangwar, A. K, Kumar, N, & Singh, H. Use of the bubaline acellular diaphragm matrix for umbilical hernioplasty in pigs. Veterinarski Arhiv (2015)., 85, 49-58.

[10] Kumar, V, Gangwar, A. K, & Kumar, N. Evaluation of the murine dermal matrix as a biological mesh in dogs. Proceedings of the National Academy of Sciences, India Section B: Biological Sciences (2016)., 86, 953-960.

[11] Vora, S. D, Kumar, V, Asodiya, F. A, Singh, V. K, Fefar, D. T, & Gajera, H. P. Bubaline diaphragm matrix: development and clinical assessment into cattle abdominal hernia repair. Brazilian Archives of Biology & Technology (2019)., 62, e19180442.

[12] Kumar, V, & Vora, S. D. Abdominal wall repair with bubaline aortic matrix in a goat. Indian Journal of Small Ruminants (2019)., 25, 255-256.

[13] Asodiya, F. A, Kumar, V, Vora, S. D, Singh, V. K, Fefar, D. T, & Gajera, H. P. Preparation, characterization, and xenotransplantation of the caprine acellular dermal matrix. Xenotransplantation (2020)., 27, e12572.

[14] Vora, S. D, Kumar, V, Singh, V. K, Fefar, D. T, & Gajera, H. P. Bubaline aortic matrix: Histologic, imaging, Fourier transform infrared

spectroscopic characterization and application into cattle abdominal hernia repair. Proceedings of the National Academy of Sciences, India Section B: Biological Sciences (2020)., 90, 161-170.

[15] Kumar, V. Abdominal intercostal hernia in a cat (*Felis domestica*). Topics in Companion Animal Medicine (2020)., 40, 100437.

[16] Gilbert, T. W, Sellaroa, T. L, & Badylak, S. F. Decellularization of tissues and organs. Biomaterials (2006)., 27, 3675-3683.

[17] Kular, J. K, Basu, S, & Sharma, R. I. The extracellular matrix: Structure, composition, age-related differences, tools for analysis and applications for tissue engineering. Journal of Tissue Engineering (2014)., 5, 1.

[18] Kumar, V, Asodiya, F. A, Singh, V. K, & Gajera, H. P. Microscopic and spectroscopic characterization of an extraskeletal intranasal osteoma in a Gir cow. Microscopy Research & Technique (2020)., 00, 1-8 http://dx.doi.org/10.1002/jemt.23613.

[19] Devarathnam, J, Sharma, A. K, Kumar, N, & Rai, R. B. Optimization of protocols for decellularization of buffalo aorta. Journal of Biomaterials and Tissue Engineering (2014)., 4, 778-785.

[20] Doyle, B. B, Bendit, E, & Blout, E. R. Infrared spectroscopy of collagen and collagen-like polypeptides. Biopolymers (1975)., 14, 937-957.

[21] Abe, Y, & Krimm, S. Normal vibrations of crystalline polyglycine II. Biopolymers (1972)., 11, 1817-1839.

[22] Payne, K. J, & Veis, A. Fourier transform IR spectroscopy of collagen and gelatin solutions: Deconvolution of the Amide I band for conformational studies. Biopolymers (1988)., 387, 1949-1960.

[23] Krimm, S, & Bandekar, J. Vibrational spectroscopy and conformation of peptides, polypeptides, and proteins. Advances in Protein Chemistry (1986)., 38, 181-364.

[24] Muyonga, J. H, Cole, C. G. B, & Duodu, K. G. Characterisation of acid soluble collagen from skins of young and adult Nile perch (*Lates niloticus*). Food Chemistry (2004)., 85, 81-89.

[25] Rey, C, Shimizu, M, Collins, B, & Glimcher, M. Resolution-enhanced Fourier transform infrared spectroscopy study of the environment of phosphate ion in the early deposits of a solid phase of calcium phosphate in bone and enamel and their evolution with age: 2. Investigations in the $\nu 3$ PO_4 domain. *Calcified Tissue International* (1991)., 49, 383-388.

[26] Thompson, T. J. U, Gauthier, M, & Islam, M. The application of a new method of Fourier transform infrared spectroscopy to the analysis of burned bone. Journal of Archaeological Science (2009)., 36, 910-914.

[27] Legan, L, Leskovar, T, Cresnar, M, Cavalli, F, Innocenti, D, & Ropret, P. Non-invasive reflection FTIR characterization of archaeological burnt bones: reference database and case studies. Journal of Cultural Heritage (2020)., 41, 13-26.

[28] Paschalis, E. P, DiCarlo, F, Betts, F, Sherman, P, Mendelsohn, R, & Boskey, A. L. FTIR microspectroscopic analysis of human osteonal bone. *Calcified Tissue International* (1996)., 59, 480-487.

[29] Rey, C, Renugopalakrishnan, V, Collins, B, & Glimcher, M. J. Fourier transform infrared spectroscopic study of the carbonate ions in bone mineral during aging. *Calcified Tissue International* (1991)., 349, 251-258.

[30] Wopenka, B, & Pasteris, J. D. A mineralogical perspective on the apatite in bone. Materials Science and Engineering: C (2005)., 25, 131-143.

[31] Paschalis, E. P, Betts, F, DiCarlo, E, Mendelsohn, R, & Boskey, A. L. FTIR microspectroscopic analysis of normal human cortical and trabecular bone. *Calcified Tissue International* (1997)., 61, 480-486.

[32] Combes, C, Cazalbou, S, & Rey, C. Apatite biominerals. Minerals (2016)., 6, 34.

[33] Akindoyo, J. O, Beg, M. D. H, Ghazali, S, & Islam, M. R. Effects of poly (dimethyl siloxane) on the water absorption and natural degradation of poly (lactic acid)/oil-palm empty-fruit-bunch fiber biocomposites. Journal of Applied Polymer Science (2015)., 132, 42784.

Chapter 5

Medical Image Classification Using the Discriminant Power Analysis (DPA) of Discrete Cosine Transform (DCT) Coefficients

Nasser Edinne Benhassine, Abdelnour Boukaache and Djalil Boudjehem

Abstract

Medical imaging systems are very important in medicine domain. They assist specialists to make the final decision about the patient's condition, and strongly help in early cancer detection. The classification of mammogram images represents a very important operation to identify whether the breast cancer is benign or malignant. In this chapter, we propose a new computer aided diagnostic (CAD) system, which is composed of three steps. In the first step, the input image is pre-processed to remove the noise and artifacts and also to separate the breast profile from the pectoral muscle. This operation is a difficult task that can affect the final decision. For this reason, a hybrid segmentation method using the seeded region growing (SRG) algorithm applied on a localized triangular region has been proposed. In the second step, we have proposed a features extraction method based on the discrete cosine transform (DCT), where the processed images of the breast profiles are transformed by the DCT where the part containing the highest energy value is selected. Then, in the feature's selection step, a new most discriminative power coefficients algorithm has been proposed to select the most significant features. In the final step of the proposed system, we have used the most known classifiers in the field of the image classification for evaluation. An effective classification has been made using the Support Vector Machines (SVM), Naive Bayes (NB), Artificial Neural Network (ANN) and *k*-Nearest Neighbors (KNN) classifiers. To evaluate the efficiency and to measure the performances of the proposed CAD system, we have selected the mini Mammographic Image Analysis Society (MIAS) database. The obtained results show the effectiveness of the proposed algorithm over others, which are recently proposed in the literature, whereas the new CAD reached an accuracy of 100%, in certain cases, with only a small set of selected features.

Keywords: medical images, breast cancer, pectoral muscle removal, discrete cosine transform, classification, SVM, KNN, ANN

1. Introduction

Cancer remains the killer disease in the world, and currently it has become a dangerous public health problem in many countries. In all kinds of cancer, the problem arises when cancer cells begin to grow in uncontrolled manner or do not die when they should do so. In addition, breast cancer is a malignant tumor that is considered the most common type of cancer occurs in women and the second type of cancer in general. It has been announced, that more than 2 million new cases have been registered worldwide in 2018 [1]. Awareness of symptoms and the need for screening are very important to reduce the risk of cancer [2].

In medical imaging, it has been shown that early detection and proper treatment of breast cancer reduces the mortality rate by 20–40% [3]. The use of Mammography, represents an effective tool in the early detection of the breast cancer. As a result, many computer-aided diagnostic (CAD) systems have been developed using digital image processing techniques applied to mammography images. These systems are very useful to help radiologists in the early detection of breast cancers and then to classify the breast tumor as malignant or benign [4–6].

In general, any CAD system can be composed of three different steps: image pre-processing step, features extraction and selection step and finally the classification step. For the breast cancer detection and classification, many works have been presented to improve the efficiency of the CAD systems. In the pre-processing step, the pectoral muscle removal and the region of interest (ROI) extraction rest a big challenge. Numerous segmentation algorithms have been also proposed to suppress the pectoral muscle [7–10]. However, there is no universal segmentation algorithm that can give acceptable results for all cases.

In the features extraction step, different techniques can be used like, shape and texture features [11, 12], morphological and texture features [13], independent component analysis (ICA) [14], the discrete cosine transform (DCT) [15], the discrete wavelet transform (DWT) [16, 17] and other transforms. In [18], the authors used non-subsampled contourlet transformation together with discrete wavelet transform with gray level co-occurrence matrix for texture features extraction. Salabat Khan et al. used a Gabor filter blank (GBF) optimized by Particle Swarm Optimization (PSO) for the extraction of Gabor characteristics [19]. Mughal B et al. used the backpropagation neural network on the hat transformation with gray level co-occurrence matrix (GLCM) features [20].

For the classification step, the most used classifiers are Artificial Neural Networks (ANN), Support Vector Machine (SVM), Naïve Bayes (NB) and *k*-Nearest Neighbors (KNN). Recently, in [21], the authors used deep learning architecture that is known as You Only Look Once (YOLO). In [22], the authors proposed a deep Convolutional Neural Network (CNN). In [23], Agnes et al., used Multiscale All Convolutional Neural Network (MA-CNN).

In this chapter, we propose a new computer-aided diagnostic system to classify breast tumors as malignant or benign. In the pre-processing step, we have proposed a new algorithm to select a limited triangular region that contains the pectoral muscle to be eliminated, and then apply the SRG segmentation algorithm. Features extraction and selection are also very important processes to improve the system performances in classification and pattern recognition methods. By using discrete Fourier transform or discrete cosine transform, we obtain a frequency domain representation of the image that can be considered as a set of features for pattern recognition problems. While the FFT give complex coefficients, the DCT provide real values in the frequency domain. We have used the DCT transform for feature extraction, and we have proposed the selection of the most significant features using the (DPA) algorithm [24]. Finally, we have evaluated the performances of the algorithm using SVM, ANN, NB and KNN classifiers and the MIAS database mammograms [25].

2. The proposed CAD system

The proposed Computer-aided diagnostic system (CAD) that is used to classify the breast tissue in mammograms as malignant or benign is divided into three basic processing steps as shown in **Figure 1**.

2.1 Image pre-processing

This step represents an important one in most CAD systems. The image pre-processing helps strongly in the selection of the region of interest (ROI) that contains the abnormalities. It is performed to remove the unwanted objects, which include artifacts, labels, background noises and to suppress the pectoral muscle (**Figure 2**). The use of efficient image processing methods is an indispensable step for achieving a high accuracy classification in CAD systems for the diagnosis of breast cancer.

2.1.1 Noise removal

There are various types of noises affected on mammogram images, such as Salt and pepper noise, Speckle noise, Gaussian noise and Poisson noise. Therefore, it is important to remove the noises to enhance the image quality on the preprocessing step. Traditionally, the median filter is a well-known used filter for this kind of noises, due to its nonlinear behavior, its simplicity and capability to preserve edges [26]. The median filter replaces each pixel value by the median of all the neighboring pixels values in a window. In this chapter, we used a (3x3) median filter to reduce noise in the mammogram images.

2.1.2 Artifacts suppression and background separation

In order to remove all unwanted objects in the selected image and separate the breast profile, we follow the next four steps:

Step (1) Thresholding of the mammogram image by 0.0706 normalized value.

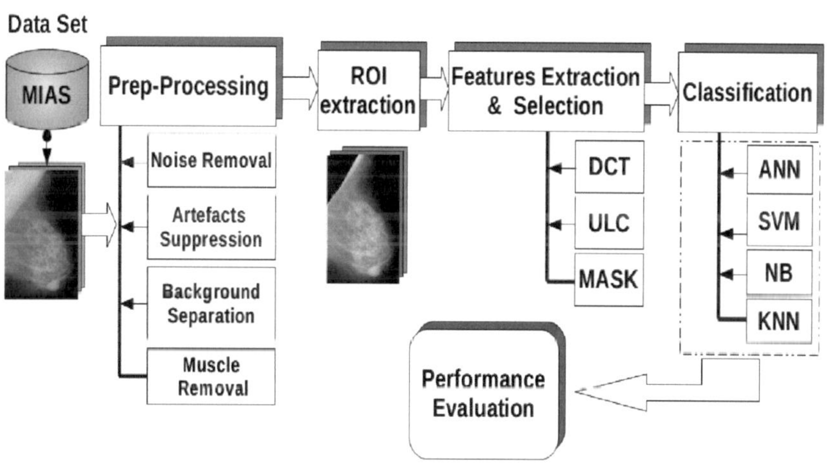

Figure 1.
Flow chart of CAD system for breast cancer classification.

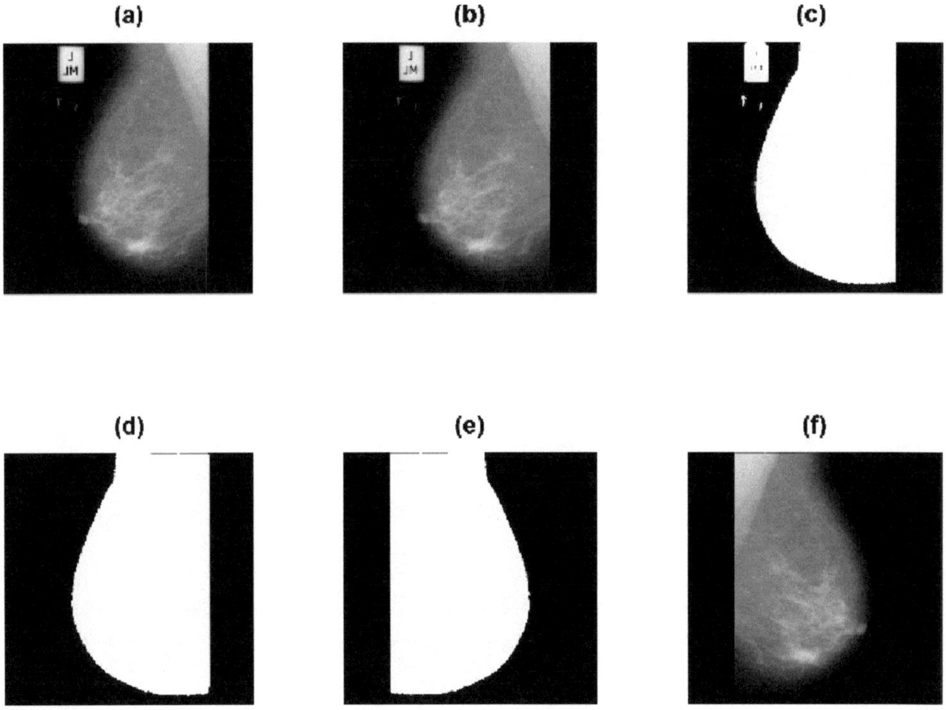

Figure 2.
The preprocessing steps of mammogram image: (a) original image, (b) noise removed image, (c) binary image, (d) largest area, (e) image right flipped, (f) parenchyma of the breast.

Step (2) Mark all regions in the thresholded image (i.e., Artifacts, labels, …).
Step (3) Calculate the area of each region, and select the largest one.
Step (4) The result of the step (3) is then used as a mask of the original grayscale mammography image.

2.1.3 Localization and extraction of the pectoral muscle

In the mammogram preprocessing, the identification and extraction of the pectoral muscle is one of the major challenges in Medio lateral Oblique (MLO) view. It could be noticed here, that this step is important to improve the diagnostic accuracy of the CAD system. The difficulty in removing the pectoral muscle is due to the following reasons [27]:

- Homogeneous area situated in the top left/right corner contains the brightest pixels in the image.

- The pectoral muscle boundary shape is concave, convex or a mixture of both of them.

- The density of the pectoral muscle area appears at approximately with similar density as the dense tissues.

- Varying position, size, shape and texture from image to image.

In this chapter, we present a new algorithm for pectoral muscle suppression, this operation is based on the Localization of the triangular region that contains the Pectoral Muscle, where the Seeded Region Growing (SRG) algorithm is invoked in this operation.

Algorithm 1 Removal of the of pectoral muscle [15]	
Input: *MLO mammograms size of 1024x1024*	
Step (1)	*All the right MLO mammograms are left–right flipped*
Step (2)	*Divided MLO into four equal quadrants of 512x512 pixels.*
Step (3)	*The upper left quadrant is cropped with removing the left background*
Step (4)	*The result is divided diagonally into two equal triangles.*
Step (5)	*Select the Seed point*
Step (6)	*Apply the SRG in the upper left triangle ABC (that contains the Pectoral Muscle)*
Step (7)	*Re-Construct the Image and removing all backgrounds*
Output:	The region of interest (ROI)

Seeded Region Growing (SRG) is a useful image segmentation technique for medical images that is initially proposed by R. Adams et *al.* [28]. This technique is robust, fast and consists of three major steps: seed selection, region growing, and region merging.

The advantage of applying the (SRG) method into the localized triangular ABC region (**Figure 3**) is to remove only the pectoral muscle, without completely suppressing the triangular region as in some other methods. **Figures 3–5** show a visual scheme of the proposed algorithm.

The Seed point is selected automatically by considering the results obtained from step (4) of algorithm 1.

Figure 5(a) shows the cropped image, where **Figure 5(b)** shows the selected region of interest (ROI). All obtained (ROI) images are resized in order to get the same dimension.

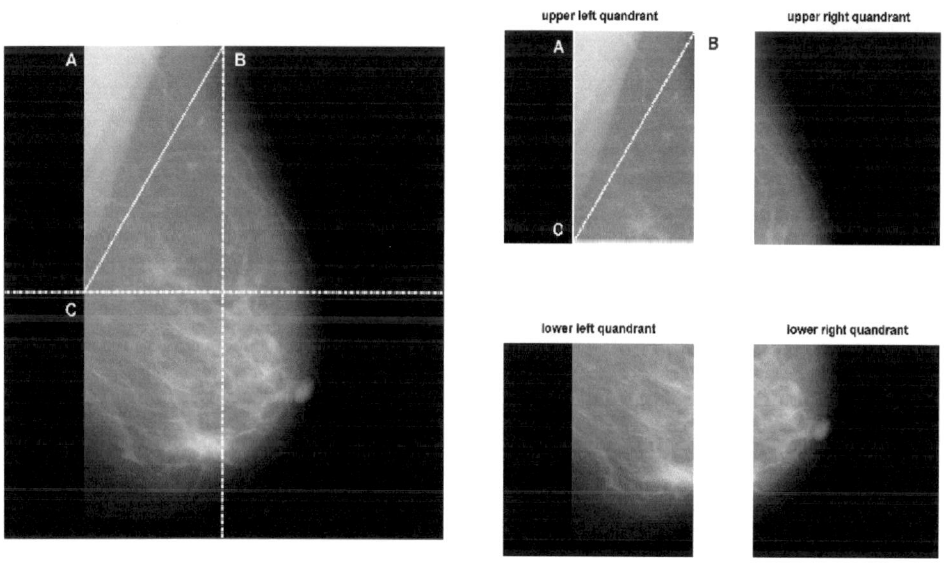

Figure 3.
Localization of the ABC triangle in the left upper quadrant.

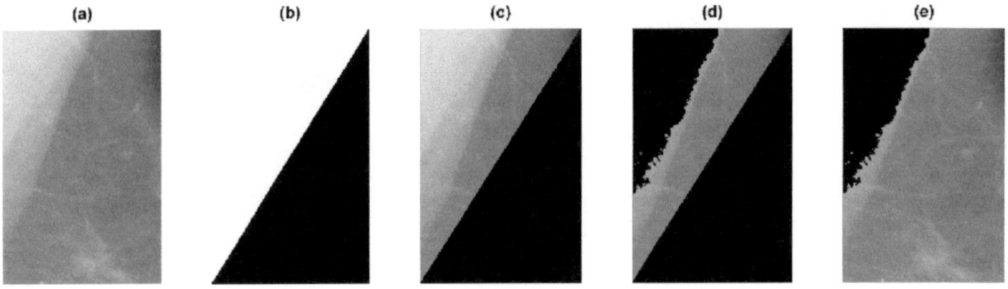

Figure 4.
Pectoral muscle segmentation steps: (a) mammogram top left quadrant, (b) triangular mask, (c) masked top left quadrant, (d) suppression of the pectoral muscle, (e) cropped top left quadrant without pectoral muscle.

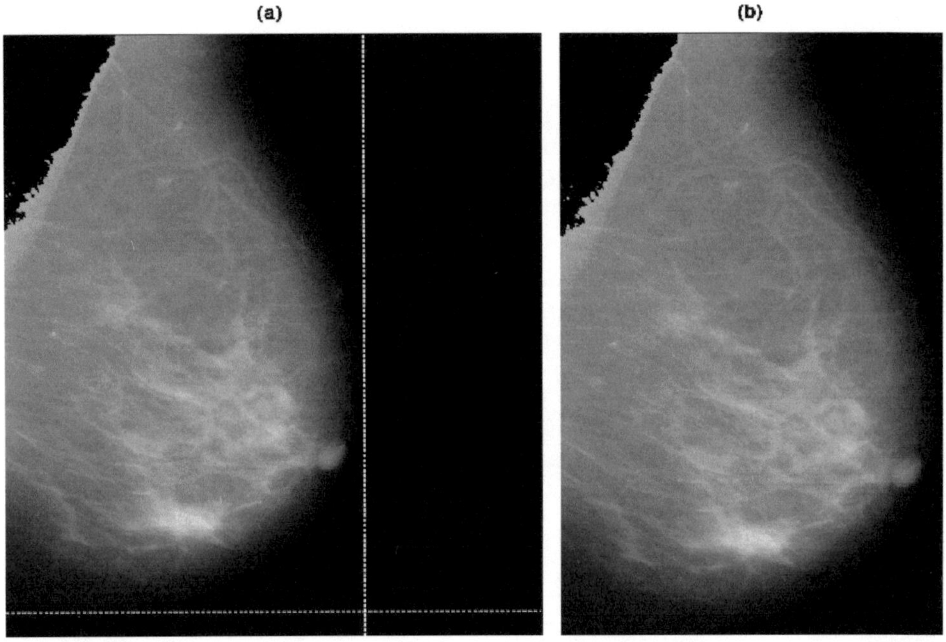

Figure 5.
Identification of the region of interest (ROI). (a) Cropped image, and (b) Selected region of Interest (ROI).

2.2 Frequency domain features extraction and selection

Features extraction plays an essential role, and a challenging step in the accurate classification and diagnostic rate of mammograms. In this chapter, we have used features extracted from the image in the frequency domain representation. The most used transform to this domain is the discrete Fourier transform with its fast algorithm (FFT) [29]. In classification problems, for example, Fourier descriptors have been used for pattern recognition [30, 31]. Another interesting transform is the discrete cosine transform (DCT) which decomposes the image on a set of cosine functions. It provides a real representation of the image contrary to the FFT, which give complex coefficients.

The frequency domain features are very used in the classification and pattern recognition field. However, the hard task is the selection of the transformed coefficients, while these coefficients do not have the same aptitude to discriminate between the different classes. However, the use of the standard approaches to select these coefficients are not always efficient in selecting the most discriminative coefficients. In this chapter, we present a novel features extraction technique that is composed of two phases. In the first one, the discrete cosine transforms (DCT) is applied on all the obtained regions of interest (ROI), and then the low frequency coefficients in the upper left corner (ULC) are retained. In the second phase, a combination of the retained frequency coefficients with the discriminative power coefficients algorithm [24] is proposed to calculate the discrimination power matrix, which is given by the ratio between the two variances, the between-class variance and the within-class variance. Where, high classification accuracies are represented by high rate values.

2.2.1 Discrete cosine transform (DCT)

This mathematical tool transforms any signal or image from the spatial domain to frequency domain. It has been widely used in digital signal and image processing, where its major advantages over the FFT reside in giving real coefficients. In addition, it concentrates the information in the low frequency region. Fast implementation can be obtained by using the FFT algorithm [29] which make the use of this transform very simple in real-time applications. It is defined by Eq. (1):

$$F(u,v) = \frac{1}{\sqrt{MN}}\alpha(u)\alpha(v)\sum_{i=0}^{M-1}\sum_{j=0}^{N-1} I(i,j)\,cos\left(\frac{(2i+1)u\pi}{2M}\right)cos\left(\frac{(2j+1)v\pi}{2N}\right) \tag{1}$$

with $0 \le u \le M$, $0 \le v \le N$, and $\alpha(u)$, $\alpha(v)$ are defined by Eq. (2), Eq. (3)

$$\alpha(u) = \begin{cases} \frac{1}{\sqrt{2}} u = 0 \\ 1\ otherwise \end{cases} \tag{2}$$

$$\alpha(v) = \begin{cases} \frac{1}{\sqrt{2}} v = 0 \\ 1\ otherwise \end{cases} \tag{3}$$

and $F(u,v)$ is the DCT coefficient matrix of the image $I(i,j)$.

After the calculation of the DCT coefficients, we retain only the 512x512 region in the upper left corner (ULC coefficients). In the feature's selection step, a new most discriminative power analysis (DPA) algorithm has been proposed to select the most significant features that have the high discrimination power (DP) values.

2.2.2 Discriminative power analysis of DCT coefficients

The calculation of the (DP) for each transformed coefficient is shown in the (DCT-DPA) algorithm shown below. Considering an image I_{ij} of size N × M. F_{uv} are the transformed coefficient by the 2D-DCT. The used database has C classes each is composed of S training images. Consequently, a total of C × S training images are used.

Input: *The number of used classes* ***C****, the training images' number* ***S*** *of each class*
The matrix Im=zeros (N, M)

Require*: Pre-processing function PPf () **and** the bi-dimensional DCT (DCT2).*

For every class c $\forall c \leq C$
  $S_{ij} = Im$;
  For every image s $(I_{ij}(S, C))$ $\forall s \leq S$
    $I_{ij}(S, C) \leftarrow PPf(I_{ij}(S, C))$ // call PPf ()
    $F_{uv}(S, C) \leftarrow DCT2(I_{ij}(S, C))$
    $S_{ij} \leftarrow S_{ij} + F_{uv}(S, C)$ //sum of images of each class
  End
  $M^c_{ij} \leftarrow S_{ij} / S$ // the mean value of each class
  $V^c_{ij} = Im$;
  For every image s $(I_{ij}(S, C))$ $\forall s \leq S$
    $I_{ij}(S, C) \leftarrow PPf(I_{ij}(S, C))$
    $F_{uv}(S, C) \leftarrow DCT2(I_{ij}(S, C))$
    $V^c_{ij} \leftarrow V^c_{ij} + (F_{uv}(S, C) - M^c_{ij})^2$ // the variance of each class
  End
End
  $M_{ij} \leftarrow \frac{1}{C}\sum_{c=1}^{C} M^c_{ij}$ // the mean of all training samples
  $V^W_{ij} \leftarrow \frac{1}{C}\sum_{c=1}^{C} V^c_{ij}$ // average the variance of all the classes
  $V^B_{ij} = Im$
For every class c $\forall c \leq C$
  For every image s $(I_{ij}(S, C))$ $\forall s \leq S$
    $I_{ij}(S, C) \leftarrow PPf(I_{ij}(S, C))$
    $F_{uv}(S, C) \leftarrow DCT2(I_{ij}(S, C))$
    $V^B_{ij} \leftarrow V^B_{ij} + (F_{uv}(S, C) - M_{ij})^2$ // the variance of the database samples
  End
End
  $Dp(i, j) = V^B_{ij} ./ V^W_{ij}$ // $1 \leq i \leq M,\ 1 \leq j \leq N$

output $Dp(i, j)$ // the discrimination power matrix

The $Dp(i,j)$ matrix is defined as the ratio of between-class variance and within-class variance. Large values of DP imply a large discrimination power of the coefficients. The selection of the DCT coefficients is made in an adaptive way according to their corresponding DP values, where the DCT coefficients with higher DP are preserved [24]. The $Dp(i,j)$ matrix is then transformed into a column vector that will be sorted in descending order. Therefore, the number k selected among the highest values, defines the number of the features to be used by the classifier. By setting the positions of selected features as ones and the others as zeros, we create a mask that can be used later in the selection process of the classification step. Alternatively, we can use a thresholding step of the DP matrix to create a mask that contain k elements as follows:

$$Mask(i,j) = \begin{cases} 1 & Dp(i,j) \geq T \\ 0 & otherwise. \end{cases} \tag{4}$$

with $1 \leq k \leq MxN$ and T is the kth highest value in the DP matrix.

2.3 Classification

The classification is the last step to identify if the breast tumor is benign or malignant. It plays a vital role in the medical image diagnosis field. Therefore, the images need to be classified with maximum accuracy. As a result, some automated classification methods have been proposed. In this part, we presented some of these classifiers including NB, SVM, ANN, and KNN that are used for breast cancer detection. A brief description of these algorithms will be presented as well as their advantages and disadvantages.

2.3.1 The support vector machine (SVM)

SVMs are a set of machine learning algorithms that help solve problems associated with classification, regression, and fault detection. They are considered to be among the algorithms that are distinguished by their strong theoretical guarantee and their great flexibility. They are also considered among the easiest algorithms in terms of ease of use even in cases where there is a little knowledge of data extraction.

The SVMs use increases widely in medical imaging field especially for breast cancer diagnosis [32]. The basic principle of SVM in this chapter is to separate and classify images into two categories malignant and benign using a hyperplane decision boundary, ensuring a maximum distance between different data sets and the boundary separating them. For linearly separable data, the hyperplane decision boundary is given by [33]:

$$g(x) = w^t x + b = \sum_{i=1}^{n} w_i^t x_i + b \tag{5}$$

where x is the input data set vector w is (n) dimensional and b is a bias. The main advantages of the (SVM) classifier can generally be listed as follows [15]:

1. SVM generally gives good accuracy with less memory use.

2. It works very well in cases where the separation margin between data sets is clear.

3. It can also solve any complex problem by specifying different kernel function.

The main disadvantages of the SVM algorithm are the difficulty of choosing the appropriate kernel function and the long training time for large datasets.

2.3.2 Artificial neural network (ANN)

The artificial neural networks are feed-forward networks that can be trained to classify inputs according to target classes. Generally, a neural network is composed of three layers: an input layer, a hidden layer and an output layer [34]. Usually, only the input and output signals of the network are already known [17]. The process of training an artificial neural network before setting it up represents a serious operation and affects directly the final obtained results. This operation depends on some constraints like the initial parameters setting, the use weights, bias and finally the used algorithm learning rate. To adjust the weights of the ANN, one can use some learning methods like the back-propagation or an optimization algorithm. In this work, the input layer is based on the number of the features selected, the hidden layer contains 10 neurons, and finally the out layer.

2.3.3 Naive Bayes classifier (NB)

Naive Bayes is becoming increasingly popular in many areas, it has shown excellent performances for classification tasks. It is a simple probabilistic classifier based on Bayes' theorem, which is based on conditional probabilities [35]. A Naive Bayes classifier assigns a new observation to the most probable class, assuming the features are conditionally independent for a given the class value. It is easy and fast to predict the class of the test data set, but their biggest disadvantage is its requirement to an independent predictor [15].

2.3.4 K-nearest neighbors (KNN)

K-nearest neighbors' classifier is a statistical non-parametric method that is used for both classification and regression [36]. In its simplest version, the *K*NN takes an arbitrary number (*k*) of neighbors nearer from the training set, and for each test point we start by determining all of its *k*-nearest neighbors among the learning points. The class that we assign to the new point is then the more frequent. In the KNN the Euclidean distance metric is used and it is given by:

$$d(x,y) = \sqrt{\sum_{i=1}^{k} (x_i - y_i)^2} \tag{6}$$

where x_i is the test sample with k features and y_i specified the training samples with k features. The Advantages of using KNN are that it is robust to noisy training data, and it is effective when used with large training data. On the other hand, it needs to determine the number of nearest neighbors (k) and the distance-based learning.

3. Results and discussions

To validate the proposed system, experiments were performed on the digital mammography images from the Mammographic Image Analysis Society (MIAS) database [25]. The MIAS database is a standard and publicly available database of digital mammogram images. Each mammogram is 1024 × 1024 pixels of size with a resolution of 200 microns. MIAS contains 322 mammograms for right and left breast of 161 patients in the mediolateral oblique (MLO) view, 61 mammograms were diagnosed as benign, 54 as malignant and 207 normal. The performance of the proposed method has been tested based on algorithms' accuracy, sensitivity and specificity using the following expressions:

$$Sensitivity = \frac{TP}{TP + FN} \times 100\ (\%) \tag{7}$$

$$Specificity = \frac{TN}{TN + FP} \times 100\ (\%) \tag{8}$$

$$Accuracy = \frac{TN + TP}{TN + FN + FP + TP} \times 100\ (\%) \tag{9}$$

Where:

TN (True Negative) is the number of a benign classified as benign.

TP (True Positive) is the number of a malignant classified as malignant.

FN (False Negative) is the number of a benign classified as malignant.

FP (False Positive) is the number of a malignant classified as benign.

In this test, we have calculated the *Dp* matrix using a set of 50 images of benign mammograms and another 50 images of malignant mammograms. These images are randomly selected as shown in the DCT-DPA algorithm. It is very important to select an equal number of samples from each class. The increase of the number of samples implies an increase of the efficiency of the performance of the classifier. However, the best results in this test are obtained using 50–50 mammogram images. Then, the database is randomly divided into train and test data sets. We have taken 113 (ROIs) divided into 70% for training and 30% for testing. The given results are the average of ten runs.

Table 1 shows a comparison of the measured performances of the SVM, ANN, NB and KNN classifiers. It is observed that the classification accuracy can reach 100% for the (ANN) classifier, is 98.8, 96.7%, 87.3% for SVM, NB and KNN

Classifiers	Sensitivity (%)	Specificity (%)	Accuracy (%)
KNN	91.05	82.67	87.3
NB	97.9	97.3	96.7
SVM	99.5	98.1	98.8
ANN	**100**	**100**	**100**

Abbreviations: ANN, artificial neural network; SVM, support vector machines; NB, Naive Bayes, KNN, K-nearest neighbors.

Table 1.
Classification performance using ANN, SVM, NB and KNN classifiers.

		ANN	SVM	NB	KNN
ULC size	N. of Features	Accuracy (%)			
128x128	60	95.6	86.76	90.6	77.9
	80	96.5	90.3	92.3	78.5
	100	98.2	94.7	92.6	79.7
256x256	60	98.2	91.2	93.2	75.0
	80	97.3	91.7	95.3	75.8
	100	98.2	93.8	93.8	76.1
512x512	60	98.2	92.1	93.8	77.0
	80	99.1	97.3	97.3	79.1
	100	**100**	**98.8**	**97.6**	**87.3**
1024x1024	60	98.2	92.6	93.1	82.6
	80	96.5	88.5	93.2	83.5
	100	97.3	91.7	94.1	85.6

Abbreviations: ANN, artificial neural network; SVM, support vector machines; NB, Naive Bayes, KNN, K-nearest neighbors.

Table 2.
Classification accuracy with various sizes of ULC and different numbers of features. Where the best results are represented by bold values.

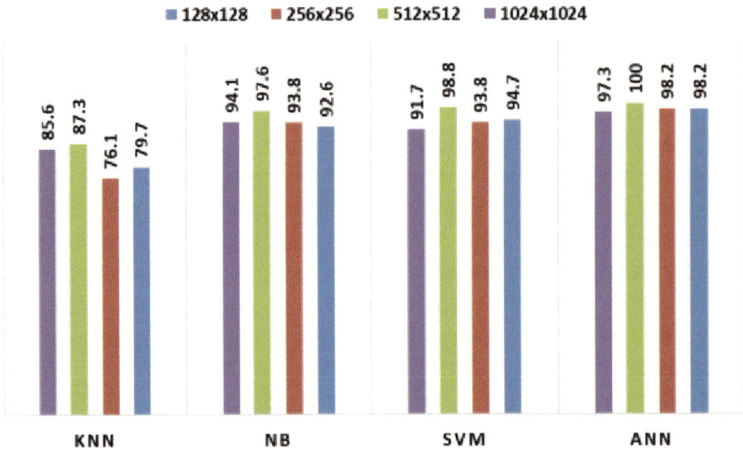

Figure 6.
Classification accuracy performances vs. the ULC sizes with 100 features.

respectively. We have evaluated the classification performances of the proposed algorithm according to the number of the used features in the classification.

The SVM, ANN, NB and KNN classifiers are used to classify input images into benign or malignant. The sensitivity, accuracy and specificity are shown in **Table 2**. According to the results in **Table 2**, we can see that small number of features (100 features in this case) can achieve best performances in the case of 512x512 ULC size. In addition, we have studied the effect of the ULC size on the obtained results. The accuracy curve of the classification accuracy versus the ULC size is shown in **Figure 6**. The number of used features is 100 features.

Figure 7 represents the variation of the classification accuracy according to different features' number with a fixed ULC size of 512x512. **Figure 8** demonstrates the classification performance of ANN using the confusion matrix for training, test and validation data.

To show the efficiency of the presented technique, **Table 3** shows a comparison between the results of the proposed algorithm with previous results, which are reported in the literature. We can see that the proposed CAD system gives better accuracy results compared to those obtained using the other methods.

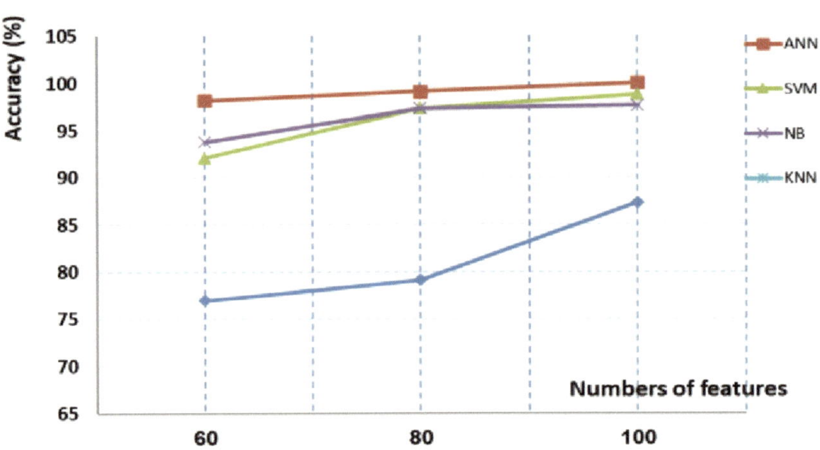

Figure 7.
Classification performances vs. the number of features. (ULC size of 512x512).

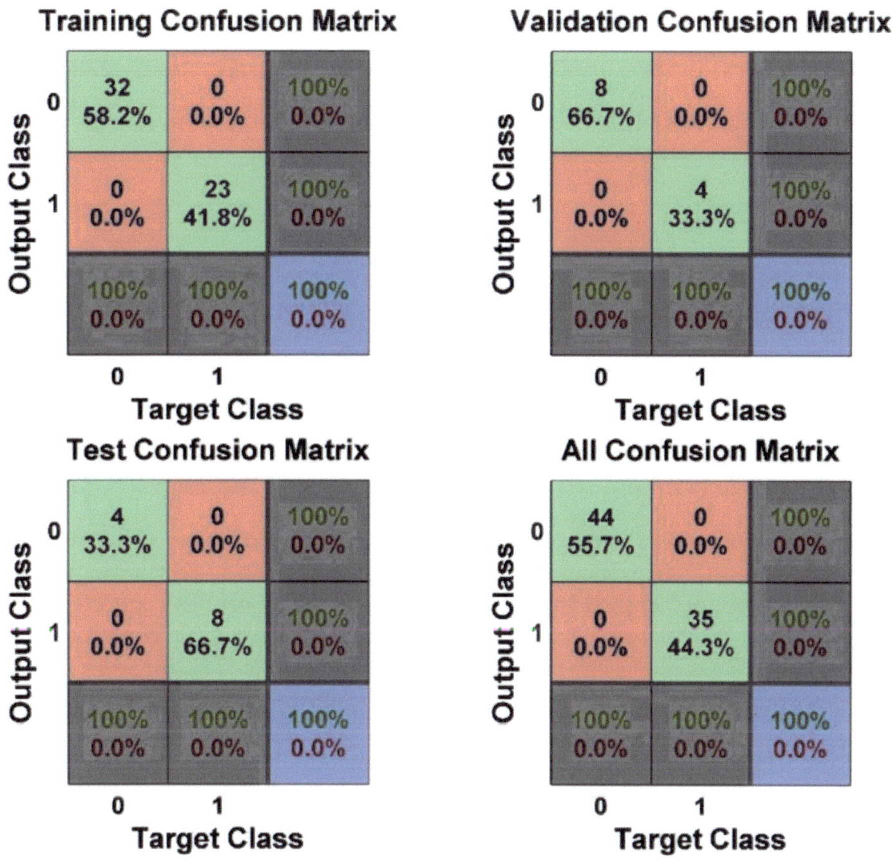

Figure 8.
Confusion matrix for training, test and validation data.

Authors	Year	Database	Classifier	Classes	Accuracy (%)
Lima [12]	2016	MIAS	SVM	2	94.1
Singh [37]	2017	MIAS	RF	2	97.3
Elmoufidi [38]	2017	MIAS	SVM	2	94.4
Mughal [20]	2018	MIAS	NNB	2 2	98.5 95
Benzebouchi [22]	2019	MIAS	SVM	2	94.0
Benhassine [15]	2019	MIAS	ANN SVM NB	2	100 94.1 92.6
El-Sokary [39]	2019	MIAS	SVM	2 3	92.5 90.0
Benhassine [17]	2020	MIAS	ANN SVM RF NB	2	99.1 99.4 98.2 97.7
Taifi [18]	2020	MIAS	SVM	2	94.1
			KNN	2	88.8

Authors	Year	Database	Classifier	Classes	Accuracy (%)
Luqman [40]	2020	MIAS CBIS-DDSM	Deep Lab Mask-RCNN	2	95.0 98.0
Proposed method	2020	MIAS	ANN SVM NB KNN	2	100 98.8 97.6 87.3

Abbreviations: ANN, artificial neural network; SVM, support vector machines; NB, Naive Bayes; RF, random Forest; KNN, k-nearest neighbors; M-RCNN and Dee Lab (two deep learning-based instance segmentation Frameworks); Subset of CBIS-DDSM Curated Breast Imaging of DDSM (Digital Database for Screening Mammography).

Table 3.
Comparison results of the proposed method with existing methods.

4. Conclusions

We have developed in the present chapter a new CAD system used for mammogram images classification. It consists of three main parts. First, we remove all unnecessary regions or objects from the input image, where we have proposed a mixed approach for pectoral muscle removing which can improve the diagnostic accuracy of the developed CAD system. Then, we have focused in our work on frequency domain features where we have used the discrete cosine transform (DCT). The extracted features are subject to a selection process that choose only the most important features. This step is done using the discriminant power analysis (DPA) algorithm. Finally, some of the most known classifiers in the field are used to make the final decision. The proposed system is evaluated on mammogram images from the MIAS database, where we have shown that a small number of selected features can give good results of the accuracy, sensitivity and specificity. The obtained results prove that the frequency domain features can give high performances especially with the use of the discrimination power analysis, and highlight the importance of DCT transform in recent artificial intelligence applications. The comparison of the obtained results with those obtained using recently proposed techniques shows the superiority of the proposed algorithm against the other methods.

Author details

Nasser Edinne Benhassine[1,2], Abdelnour Boukaache[1] and Djalil Boudjehem[1*]

1 Advanced Control Laboratory (LABCAV), University of 8th May 1945, Guelma, Algeria

2 University center of Aflou, Algeria

*Address all correspondence to: boudjehem.djalil@gmail.com

References

[1] The World Health Organization (WHO): http://www.who.int/mediacentre/factsheets/fs297/en/ [Accessed: 2018-10-28]

[2] American Cancer Society, Breast Cancer Facts & Figures 2017–2018, Atlanta: *American Cancer Society, Inc*. 2017.

[3] Jalalian A, Mashohor S, Mahmud R, Karasfi B, Saripan MI, Ramli AR. Computer-assisted diagnosis system for breast cancer in computed tomography laser mammography (CTLM). J Digit Imaging. 2017; 30:796–811

[4] Pisano ED, Shtern F. Image processing and computer aided diagnosis in digital mammography: a clinical perspective. Int J Pattern Recognit Artif Intell. 1993; 7(06):1493–1503.

[5] Giger ML, Karssemeijer N, Armato SG. Computer-aided diagnosis in medical imaging. IEEE Trans Med Imaging. 2001;20(12):1205–1208.

[6] Vyborny CJ, Giger ML, Nishikawa RM. Computer-aided detection and diagnosis of breast cancer. Radiol Clin North Am. 2000;38(4):725–740.

[7] Singh VP, Srivastava R. Content-based mammogram retrieval using wavelet based complete LBP and K-means clustering for the diagnosis of breast cancer. Int J Hybrid Intell Syst. 2017; 14:31–39.

[8] Karssemeijer N. Automated classification of parenchymal patterns in mammograms. Phys Med Biol. 1998;43 (2):365–378.

[9] Xu W, Xia S, Duan H. A novel computer-aided diagnosis system of the mammograms. In: Huang DS, Li K, Irwin KGW, eds. Intelligent Computing in Signal Processing and Pattern Recognition. 2006;639–644.

[10] Nagi J, Kareem SA, Nagi F, Ahmed SK. Automated breast profile segmentation for ROI detection using digital mammograms. Biomedical Engineering and Sciences (IECBES); November 30–December 2, 2010; Kuala Lumpur, Malaysia: IEEE; 2010. p. 87–92

[11] Zakeri FS, Behnam H, Ahmadinejad N. Classification of benign and malignant breast masses based on shape and texture features in sonography images. J Med Syst. 2012;36 (3):1621–1627.

[12] de Lima SM, da Silva-Filho AG, dos Santos WP. Detection and classification of masses in mammographic images in a multi-kernel approach. Comput Methods Programs Biomed. 2016; 134: 11–29.

[13] Oliver A, Freixenet J, Marti R, et al. A novel breast tissue density classification methodology. IEEE Trans Inf Technol Biomed. 2008;12(1):55–65.

[14] Campos LFA, Silva AC, Barros AK. Diagnosis of breast cancer in digital mammograms using independent component analysis and neural networks, vol 3773. Iberoamerican Congress on Pattern Recognition; November 15–18, 2005; Havana, Cuba; 2005:160–169.

[15] Benhassine, N.E., Boukaache, A. and Boudjehem D. Classification of mammogram images using the energy probability in frequency domain and most discriminative power coefficients. *Int J Imaging Syst Technol*. 30:45–56, 2019. https://doi.org/10.1002/ima.22352.

[16] Lahmiri S, Boukadoum M. Hybrid discrete wavelet transform and Gabor

filter banks processing for features extraction from biomedical images. J Med Eng. 2013; 2013:1–13.

[17] Benhassine, N.E., Boukaache, A. and Boudjehem D. A new CAD System for Breast Cancer Classification using Discrimination Power Analysis of Wavelet's Coefficients and Support Vector Machine. *Journal of Mechanics in Medicine and Biology*. https://doi.org/10.1142/S0219519420500360.

[18] Taifi K, et al. Mammogram Classification Using Nonsubsampled Contourlet Transform and Gray-Level Co-Occurrence Matrix. *Crit Approaches to Inf Retr Res Glob*. 2020:239–55.

[19] Khan S, Hussain M, Aboalsamh H, Mathkour H, Bebis G, Zakariah M. Optimized Gabor features for mass classification in mammography. Appl Soft Comput. 2016; 44:267–280.

[20] Mughal B, Sharif M, Muhammad N, Saba T. A novel classification scheme to decline the mortality rate among women due to breast tumor. Microsc Res Tech. 2018. https://doi.org/10.1002/jemt.22961.

[21] Al-masni MA, et al. Simultaneous detection and classification of breast masses in digital mammograms via a deep learning YOLO-based CAD system. *Comput Methods Programs Biomed*. 2018; 157:85–94

[22] Benzebouchi NE, Azizi N, Ayadi K A. computer-aided diagnosis system for breast cancer using deep convolutional neural networks. In: Computational intelligence in data mining Springer, Singapore. Advances in intelligent systems and computing. 2019; 711.

[23] Agnes, S. Akila, J. Anitha, S. Immanuel Alex Pandian, and JDP. Classification of Mammogram Images Using Multiscale all Convolutional Neural Network (MA-CNN). J Med Syst. 2020; 44:30.

[24] Dabbaghchian S, Ghaemmaghami MP, Aghagolzadeh A. Feature extraction using discrete cosine transform and discrimination power analysis with a face recognition technology. Pattern Recognition. 2010; 43(4):1431–1440.

[25] Suckling J, Parker J, Dance D, et al. Mammographic Image Analysis Society (MIAS) database v1.21; 2015.

[26] Kshema J, George M, Dhas DAS. Preprocessing filters for mammogram images: a review. IEEE Conference on Emerging Devices and Smart Systems; Mar 3–4, 2017; Tamilnadu, India; 2017: 3–4.

[27] Sapate S, Talbar S. An overview of pectoral muscle extraction algorithms applied to digital mammograms. In: Dey N, Bhateja V, Hassanien A, eds. Medical Imaging in Clinical Applications. Cham: Springer. 2016:19–54.

[28] Adams R, Bischof L. Seeded region growing. IEEE Trans Pattern Anal Mach Intell. 1994;16(6):641–647.

[29] Cooley, J. W., & Tukey, J. W. (1965). An algorithm for the machine calculation of complex Fourier series. Mathematics of computation, 19(90), 297–301.

[30] Lu, Y., Schlosser, S., & Janeczko, M. (1993). Fourier descriptors and handwritten digit recognition. Machine Vision and Applications, 6(1), 25–34.

[31] Hamd, M. H., & Ahmed, S. K. (2017). Fourier descriptors for iris recognition. International Journal of Computing and Digital Systems, 6(05), 285–291.

[32] Wang C, Mai X, Lin G, Tan Kuo C. Classification for breast MRI using support vector machine. International Conferences on Computer and Information Technology Workshops; 8–

11 July 2008; Sydney, Australia: IEEE; 2008: p. 362–367

[33] Vapnik V. Statistical Learning Theory. New York: Wiley; 1998.

[34] Singh S, Gupta PR. Breast cancer detection and classification using neural network. Int J Adv Eng Sci Technol. 2011; 6:4–9.

[35] Tang B, Kay S, He H, Toward. optimal feature selection in naive Bayes for text categorization. IEEE Trans Knowl Data Eng. 2016; 28:2508–2521.

[36] Altman, Naomi S. An introduction to kernel and nearest-neighbour non parametric regression. The American Statistician.1992;46(3):175–185. doi: 10.1080/00031305.1992.10475879.

[37] Singh VP, Srivastava S, Srivastava R. Effective mammogram classification based on center symmetric-LBP features in wavelet domain using random forest classifier. Technol Health Care. 2017; 1: 1–19.

[38] Elmoufidi A, et al. Automatic Detection of Suspicious Lesions in Digital X-ray Mammograms. In : El-Azouzi R, Menasche D, Sabir E, De Pellegrini F, Benjillali M, eds. Advances in Ubiquitous Networking 2. UNet 2016. Lecture Notes in Electrical Engineering, Singapore: Springer. 2017; 397.

[39] N. El-Sokary, et al. Computer-Aided Detection System for Breast Cancer Based on GMM and SVM. Arab Journal of Nuclear Sciences and Applications. 2019; 52(2):142–150.

[40] Luqman Ahmed et AL, Images data practices for Semantic Segmentation of Breast Cancer using Deep Neural Network. *Journal of Ambient Intelligence and Humanized Computing*. 2020. https://doi.org/10.1007/s12652-020-01680-1.

Chapter 6

Path Integral Two Dimensional Models of P– and D–Wave Superconductors and Collective Modes

Peter Brusov and Tatiana Filatova

Abstract

The main parameter, which describes superfluids and superconductors and all their main properties is the order parameter. After discovery the high temperature superconductors (HTSC) and heavy fermion superconductors (HFSC) the unconventional pairing in different superconductors is studied very intensively. The main problem here is the type of pairing: singlet or triplet, orbital moment of Cooper pair value L, symmetry of the order parameter etc. Recent experiments in Sr_2RuO_4 renewed interest in the problem of the symmetry of the order parameters of the HTSC. The existence of CuO_2 planes – the common structural factor of HTSC – suggests we consider two-dimensional (2D) models. A 2D– model of *p*–pairing using a path integration technique has been developed by Brusov and Popov. A 2D model of *d*–pairing within the same technique has been developed by Brusov et al. All properties of 2D–superconductors (for example, of CuO_2 planes of HTSC) and, in particular, the collective excitations spectrum, are determined by these functionals. We consider all superconducting states, arising in symmetry classification of p-wave and d-wave 2D–superconductors, and calculate the full collective modes spectrum for each of these states. This will help to identify the type of pairing and the symmetry of the order parameter in HTSC and HFSC.

Keywords: path integral, two-dimensional models, P- and D-wave superconductors, collective modes

1. Introduction

The main parameter, which describes superfluids and superconductors and all their main properties is the order parameter, which is equal to zero above transition temperature T_c into superconducting (superfluid) state and becomes nonzero below T_c. In Bose–systems the transition is caused by Bose–Einstein condensation of bosons while in case of Fermi–systems first the pairing of fermions with creation of Bose particles (Cooper pairs) takes place with their subsequent condensation. Besides the ordinary superconductors (where traditional s–pairing takes place) after discovery the high temperature superconductors (HTSC) and heavy fermion superconductors (HFSC) the unconventional pairing in different superconductors is studied very intensively [1–5]. The main problem here is the type of pairing: singlet or triplet, orbital moment of Cooper pair value L, symmetry of the order parameter etc.

Recent experiments in Sr_2RuO_4 [2–4] renewed interest in the problem of the symmetry of superconducting order parameters of the high temperature superconductors (HTSC).

Sr_2RuO_4 has been the candidate for a spin–triplet superconductor for more than 25 years. Recent NMR experiments have cast doubt on this candidacy. Symmetry–based experiments are needed that can rule out broad classes of possible superconducting order parameters. In Ref. 3 authors use the resonant ultrasound spectroscopy to measure the entire symmetry–resolved elastic tensor of Sr_2RuO_4 through the superconducting transition. They observe a thermodynamic discontinuity in the shear elastic modulus c_{66}, which implies that the superconducting order parameter has two components. A two–component p–wave order parameter, such as $p_x + ip_y$, satisfies this requirement. As this order parameter appears to have been precluded by recent NMR experiments, the alternative two–component order parameters of Sr_2RuO_4 are as following $\{d_{xz}, d_{yz}\}$ and $\{d_{x2-y2}, g_{xy(x2-y2)}\}$.

Authors of Ref. 4 have come to similar conclusions. They use ultrasound velocity to probe the superconducting state of Sr_2RuO_4. This thermodynamic probe is sensitive to the symmetry of the superconducting order parameter. Authors observe a sharp jump in the shear elastic constant c_{66} as the temperature is increased across the superconducting transition. This supposes that the superconducting order parameter is of a two–component nature.

The existence of CuO_2 planes [6] – the common structural factor of HTSC – suggests we consider 2D models. A 2D– model of *p*–pairing using a path integration technique has been developed by Brusov and Popov [7, 8]. A 2D model of *d*–pairing within the same technique has been developed by Brusov et al. [9–14]. The models use the hydrodynamic action functionals, which have been obtained by path integration over “fast” and “slow” Fermi–fields. All properties of 2D–superconductors (for example, of CuO_2 planes of HTSC) and, in particular, the collective excitations spectrum, are determined by these functionals. We consider all superconducting states, arising in symmetry classification of 2D–superconductors and calculate the full collective modes spectrum for each of these states. Current study continue our previous investigation [5], where we consider the problem of distinguish the mixture of two d–wave states from pure d–wave state of HTSC.

2. Two–dimensional models of p– and d–pairing in unconventional superconductors

2.1 *p*–Pairing

Below we develop 2D–model of *p*–pairing starting with the 3D scheme considered by Brusov et al. [9–14].

Two main distinctions between 3D–case and 2D–case are as follows:

a. The Cooper pair orbital moment $\boldsymbol{l}$ ($\boldsymbol{l} = 1$) should be perpendicular to the plane and can have only two projections on the $\hat{z}$–axis: ± 1. The *p*– pairing is a triplet, thus the total spin of the pair is equal to 1, and in the case of 2D *p*–pairing we have $3 \times 2 \times 2 = 12$ degrees of freedom. In this case one can describe the superconducting state by complex 2×3 matrices $c_{ia}(p)$. The number of the collective modes in each phase is equal to the number of degrees of freedom. Just remind that in the 3D case this number is equal to 18.

b. Vector $\mathbf{x}$ is a 2D vector and square “volume” will be $S = L^2$ (instead of $V = L^3$ in 3D case).

2.1.1 Two–dimensional p*–wave superconducting states*

Effective action In case of two–dimensional p–wave superconductivity effective action takes a form (see the case of two–dimensional superfluidity of ^{3}He in Chapter XIX of Ref. 1)

$$S_{eff} = -\beta V \frac{16\pi^2 T_C \Delta T}{7\zeta(3)} F \tag{1}$$

where

$$\begin{aligned} F = -trAA^{+} + \nu trA^{+}AP + \left(trA^{+}A\right)^{2} + trAA^{+}AA^{+} + trAA^{+}A^{*}A^{T} - \\ -trAA^{T}A^{*}A^{+} - (1/2)trAA^{T}trA^{*}A^{+}, \\ \nu = 7\zeta(3)\mu^2H^2/4\pi^2T_C\Delta T \end{aligned} \tag{2}$$

The effective action F is identical in form with that arising in the case of three–dimensional (3D) superconducting system. The difference is connected with the fact that the matrix A with elements a_{ia} for the two–dimensional system is a 2×3 matrix instead of 3×3 matrix in the case of three–dimensional (3D) superconducting system. The matrix P is the projector on the third axis: $P = \delta_{i3}\delta_{j3}$

$$P = \begin{pmatrix} 0 & 0 & 0 \\ 0 & 0 & 0 \\ 0 & 0 & 1 \end{pmatrix}$$

The following equation for the condensate matrix A could be obtained by minimizing F:

$$\begin{aligned} -A + \nu AP + 2(trAA^{+})A + 2AA^{+}A + 2A^{*}A^{T}A - 2AA^{T}A^{*} - \\ -\left(trAA^{T}\right)A^{*} = 0. \end{aligned} \tag{3}$$

There are several solutions of Eq. (3), corresponding to the different superfluid phases. Let us consider the following possibilities:

$$\begin{aligned} A_1 = \frac{1}{2}\begin{pmatrix} 1 & 0 & 0 \\ i & 0 & 0 \end{pmatrix}, A_2 = \frac{1}{2}\begin{pmatrix} 1 & 0 & 0 \\ 0 & 1 & 0 \end{pmatrix}, A_3 = \frac{1}{4}\begin{pmatrix} 1 & i & 0 \\ i & -1 & 0 \end{pmatrix}, \\ A_5 = \frac{1}{\sqrt{3}}\begin{pmatrix} 0 & 0 & 0 \\ 0 & 1 & 0 \end{pmatrix}, A_6 = \frac{1}{2}\begin{pmatrix} 1 & 0 & 0 \\ 0 & -1 & 0 \end{pmatrix}, A_7 = \frac{1}{2}\begin{pmatrix} 0 & +1 & 0 \\ 1 & 0 & 0 \end{pmatrix}, \\ A_8 = \left(\frac{1-\nu}{3}\right)^{1/2}\begin{pmatrix} 0 & 0 & 1 \\ 0 & 0 & 0 \end{pmatrix}, A_9 = \left(\frac{1-\nu}{4}\right)^{1/2}\begin{pmatrix} 0 & 0 & 1 \\ 0 & 0 & i \end{pmatrix}. \end{aligned} \tag{4}$$

The corresponding values of the effective action F are equal to:

$$\begin{aligned} F_1 = -\frac{1}{4}, F_2 = -\frac{1}{4}, F_3 = -\frac{1}{8}, F_4 = -\frac{1}{6}, F_5 = -\frac{1}{6}, \\ F_6 = -\frac{1}{4}, F_7 = -\frac{1}{4}, F_8 = -(1-\nu)^2/6, F_9 = -(1-\nu)^2/4. \end{aligned} \tag{5}$$

The quantity of the effective action F for the first eight phases does not depend on H. The minimum value of $F = -1/4$ is reached for phases A_1 and A_2 as well as for the phases with matrices A_6 and A_7 and A_9 (last state has minimum energy in zero magnetic field ($\nu = 0$)).

The first two phases have been discovered by Brusov and Popov [7, 8] in the films of superfluid ^{3}He. Authors [7, 8] have called them the a – and b –phases and have proved that the phases a– and b– are stabile relative to the small perturbations. Brusov and Popov [7, 8] have calculated the full collective mode spectrum for two these phases. Brusov et al. [9–14] have calculated the full collective mode spectrum for A_6 and A_7 states.

2.1.2 The collective mode spectrum

The full collective mode spectrum for each of these phases consists of 12 modes (the number of degrees of freedom). Among them we have found Goldstone modes as well as high frequency modes (with energy (frequency) which is proportional to energy of the gap in single–particle spectrum).

The results obtained by Brusov and Popov [7, 8] and Brusov et al. [9–14] are shown below for collective mode spectrum for different two–dimensional superconducting states under p–pairing.

The collective mode spectrum for a–phase with order parameter $\frac{1}{2}\begin{pmatrix} 1 & 0 & 0 \\ i & 0 & 0 \end{pmatrix}$:

$$E^2 = \frac{c_F^2 k^2}{2}\left(1 - \frac{5c_F^2 k^2}{96\Delta^2}\right), (3\ \text{modes})$$
$$E^2 = 2\Delta^2 + c_F^2 k^2/2, (6\ \text{modes}) \tag{6}$$
$$E^2 = 4\Delta^2 + (0.500 + i0.433)c_F^2 k^2.(3\ \text{modes})$$

The collective mode spectrum for b–phase with order parameter $\frac{1}{2}\begin{pmatrix} 1 & 0 & 0 \\ 0 & 1 & 0 \end{pmatrix}$:

$$E^2 = \frac{c_F^2 k^2}{2}\left(1 - \frac{5c_F^2 k^2}{48\Delta^2}\right), (2\ \text{modes})$$
$$E^2 = \frac{3c_F^2 k^2}{4}\left(1 - \frac{c_F^2 k^2}{72\Delta^2}\right), (1\ \text{mode})$$
$$E^2 = \frac{c_F^2 k^2}{4}\left(1 - \frac{c_F^2 k^2}{48\Delta^2}\right), (1\ \text{mode})$$
$$E^2 = 2\Delta^2 + c_F^2 k^2/2, (4\ \text{modes}) \tag{7}$$
$$E^2 = 4\Delta^2 + (0.500 - i0.433)c_F^2 k^2, (2\ \text{mode})$$
$$E^2 = 4\Delta^2 + (0.152 - i0.218)c_F^2 k^2, (1\ \text{mode})$$
$$E^2 = 4\Delta^2 + (0.849 - i0.216)c_F^2 k^2.(1\ \text{mode})$$

It is seen that in a– and b–phases the so–called two–dimensional (2D) sound with velocity $v_2 = c_F/\sqrt{2}$ exists. Note that dispersion coefficient of 2D–sound in b–phase is twice higher than in a–phase. We should remind that in bulk systems the three–dimensional sound with velocity $v_3 = c_F/\sqrt{3}$ is well known). After Brusov et al. [7, 8] this result has been reproduced by a number of authors (Nagai [15], Tewordt [16] etc.).

The collective mode spectrum for the phase with order parameter $\frac{1}{2}\begin{pmatrix} 0 & 0 & 1 \\ 0 & 0 & i \end{pmatrix}$:

$$\begin{aligned} E^2 &= 0, (3\text{ modes}); \\ E^2 &= 2\Delta^2, (6\text{ modes}); \\ E^2 &= 4\Delta^2. (3\text{ modes}) \end{aligned} \tag{8}$$

The collective mode spectrum for two phases with order parameters $\frac{1}{2}\begin{pmatrix} 0 & \pm 1 & 0 \\ 1 & 0 & 0 \end{pmatrix}$:

$$\begin{aligned} E^2 &= 0, (4\text{ modes}); \\ E^2 &= 2\Delta^2, (4\text{ modes}); \\ E^2 &= 4\Delta^2. (4\text{ modes}) \end{aligned} \tag{9}$$

The collective mode spectrum for the phase with order parameter $\frac{1}{2}\begin{pmatrix} 1 & 0 & 0 \\ 0 & -1 & 0 \end{pmatrix}$:

$$\begin{aligned} E^2 &= 0, (4\text{ modes}); \\ E^2 &= 2\Delta^2, (4\text{ modes}); \\ E^2 &= 4\Delta^2. (4\text{ modes}) \end{aligned} \tag{10}$$

3. Two–dimensional d–Wave superconductivity

3.1 2D–model of d–pairing in CuO_2 planes of HTSC

The existence of CuO_2 planes — the common structural factor of HTSC — suggests we consider two–dimensional (2D) models. For two–dimensional (2D) quantum antiferromagnet (AF) it was shown that only the *d*–channel provides an attractive interaction between fermions. The *d*– pairing arises also in symmetry classifications of CuO_2 planes HTSC. In Sr_2RuO_4 where the p–pairing appears to have been precluded by recent NMR experiments, the two–component d–wave order parameters, namely $\{d_{xz}, d_{yz}\}$ and even with admixture of g–wave $\{d_{x2-y2}, g_{xy(x2-y2)}\}$, are now the prime candidates for the order parameter of the quasi–two–dimensional Sr_2RuO_4.

The two–dimensional (2D) model of *d*– pairing in the CuO_2 planes of HTSC has been developed by Brusov and Brusova (BB) [9, 10, 13] and Brusov, Brusova and Brusov (BBB) [14] using a path integration technique. The hydrodynamic action functional, obtained by path integration over "fast" and "slow" Fermi–fields, has been used under construction of this model. This hydrodynamic action functional determines all properties of the CuO_2 planes and, in particular, the spectrum of collective excitations.

To develop the model of d–pairing in the two–dimensional (2D)–case we modify the three–dimensional (3D) considered by us in Ref. 1.

The main distinctions between 3D and 2D cases are as follows:

a. The orbital moment $\vec{l}\left(\left|\vec{l}\right|=2\right)$ should be perpendicular to the plane and can have only two projections on the $\hat{z}$– axis: ± 2 instead of the three–dimensional (3D) case where the orbital moment can have five projections on the $\hat{z}$– axis: ± 2; ± 1;0. Because the d– pairing is a singlet the total spin of the pair is equal zero, so in the case of the two–dimensional (2D) d– pairing one has $1\times 2\times 2=4$ degrees of freedom. Thus the superconductive state in this case can be described by complex symmetric traceless 2×2 matrices $c_{ia}(p)$, which have the same number of degrees of freedom $(2\times 2\times 2-2-2=4)$. This number is equal to the number of the CM in each phase. Note that in the three–dimensional (3D) case this number is equal to 10, as well as the number of the collective modes in each phase.

b. The pairing potential t is given by:

$$t=v\left(\hat{k},\hat{k}'\right)=\sum_{m=-2,2} g_m Y_{2m}\left(\hat{k}\right)Y^*_{2m}\left(\hat{k}'\right) \tag{11}$$

We consider the case of circular symmetry $g_2=g_{-2}=g$, which is describes by one coupling constant g. Note, that less symmetric cases require both constants g_2 and g_{-2}. We consider the circularly symmetric case where:

$$v\left(\hat{k},\hat{k}'\right)=g\left[Y_{2-2}\left(\hat{k}\right)Y^*_{2-2}\left(\hat{k}'\right)+Y_{22}\left(\hat{k}\right)Y^*_{22}\left(\hat{k}'\right)\right] \tag{12}$$

c. $\mathbf{x}$ will be a 2D–vector and square "volume" will be $S=L^2$ (instead of $V=L^3$ as in 3D case).

Account these distinctions between the two–dimensional (2D) and the three–dimensional (3D) cases we will describe our Fermi–system by the anticommuting functions $\chi_s(\mathbf{x},\tau)$, $\overline{\chi}_s(\mathbf{x},\tau)$, defined in the square volume $S=L^2$ and antiperiodic in "time" τ with period $\beta=T^{-1}$.

After path integrating over slow and fast Fermi–fields (which is a very similar to 3D one) one gets the effective action functional S_{eff}, which takes (formally) the same form as in 3D case.

The number of degrees of freedom in the case of two–dimensional (2D) d–pairing is equal to 4. By the other words, one has two complex canonical variables. It is easy to see from non–diagonal elements of $\hat{M}$ matrix that the following canonical variables should be chosen:

$$c_1=c_{11}-c_{22}, c_2=c_{12}+c_{21}. \tag{13}$$

One has for the conjugate variables:

$$c_1^+=c_{11}^+-c_{22}^+, c_2^+=c_{12}^++c_{21}^+. \tag{14}$$

Below we transform the effective action functional S_{eff} to these new variables. One has:

$$S_{\text{eff}}=(2g)^{-1}\sum_{p,j}c_j^+(p)c_j(p)+\frac{1}{2}\ln\det\frac{\hat{M}\left(c_j^+,c_j\right)}{\hat{M}\left(c_j^{+(0)},c_j^{(0)}\right)} \tag{15}$$

where

$$\begin{aligned} M_{11} &= Z^{-1}[i\omega - \xi + \mu(\mathbf{H}\boldsymbol{\sigma})]\delta_{p_1p_2} \\ M_{22} &= Z^{-1}[-i\omega + \xi + \mu(\mathbf{H}\boldsymbol{\sigma})]\delta_{p_1p_2} \\ M_{12} = M_{21}^{+} &= \sigma_0\alpha(\beta S)^{-1/2}(c_1\cos 2\phi + c_2\sin 2\phi). \end{aligned} \tag{16}$$

The effective functional S_{eff} determines all properties of considering model system – superconducting CuO_2 planes. It determines, in particular, the full collective–mode spectrum, consisting of four collective–modes in each phase.

3.2 The collective mode spectrum

Two SC states arise in the symmetry classification of CuO_2 planes with OP which are proportional to $\begin{pmatrix} 1 & 0 \\ 0 & -1 \end{pmatrix}$ and $\begin{pmatrix} 0 & 1 \\ 1 & 0 \end{pmatrix}$ respectively. In the former phase the gap is proportional to $Y_{22} + Y_{2-2} \sim \sin^2\theta|\cos 2\phi| \sim |\cos 2\phi|$ while in the later one is proportional to $-i(Y_{22} - Y_{2-2}) \sim \sin^2\theta|\sin 2\phi| \sim |\sin 2\phi|$. For 2D case we put $\theta = \pi/2$ and $\sin\theta = 1$.

Brusov and Brusova [9, 10] and Brusov, Brusova and Brusov [14] have calculated the collective– mode spectrum for both of these states. In the first approximation the collective excitations spectrum is determined by the quadratic part of S_{eff}, obtained by the shift $c_j(p) \to c_j^{(0)} + c_j(p)$ in S_{eff}. Here $c_j^{(0)}$ are the condensate values of the canonical Bose–fields $c_j(p)$.

The collective mode spectrum for the phases with order parameters $\begin{pmatrix} 1 & 0 \\ 0 & -1 \end{pmatrix}$ **and** $\begin{pmatrix} 0 & 1 \\ 1 & 0 \end{pmatrix}$.

The spectra in both phases turns out to be identical. Brusov and Brusova [9, 10] found two high frequency modes in each phase with following energies (frequencies):

$$\begin{aligned} E_1 &= \Delta_0(1.42 - i0.65), \\ E_2 &= \Delta_0(1.74 - i0.41). \end{aligned} \tag{17}$$

Note that the energies of both modes turn out to be complex. This results from the *d*–pairing, or in other words, via the disappearance of a gap in the chosen directions. In this case the Bose–excitations decay into fermions. This leads to a damping of the collective modes. The value of imaginary part of energy is 23% for the second mode and 46% for first one. Thus both modes should be regarded as resonances and the second mode is better defined than the first.

The other two modes are Goldstone or low–energy modes (with energy $\leq 0.1\Delta_0$).

4. Lattice symmetry and collective mode spectrum

The made calculations of the collective mode spectrum are not completely self-–consistent, because Brusov, Brusova and Brusov [14] working within spherical symmetry approximation, use the order parameters obtained with taking the lattice symmetry into account. The taking the lattice symmetry into account, as we mentioned above, requires a few coupling constants using instead of one. The number of collective excitations (collective modes) in superconducting state, which is equal to number of degrees of freedom, will change too (note, that in case of spherical symmetry it is equal to 10). In case of the simple irreducible representation (IR) the number of collective modes is equal to twice number of irreducible representation dimensionality. For orthorhombic (OR) symmetry and singlet pairing all irreducible representations are one dimensional (1D), so in each superconducting state there are two modes corresponding to phase and amplitude variations. Amplitude mode is high frequency with $E \approx 2\Delta$, where Δ is the gap in a single particle spectrum.

Among irreducible representations of tetragonal (TG) symmetry there are 1D as 2D (remind that we consider the singlet pairing). Thus in addition to the superconducting states, with two collective modes of conventional superconductors there are states which have four collective modes, none of which are Goldstone. We would like to mention, that for cylindrical Fermi–surface (D_∞) among collective modes there is Goldstone mode in $(1, 0)$ and $(1, 1)$ states but there is not Goldstone mode in $(1, i)$ state.

Because it looks like that there is a mixture of different irreducible representations (corresponding, for example, to *s*– and *d*–wave states or to two different *d*–wave states: $d_{x^2-y^2}$ and d_{xy}; or d_{xz} and d_{yz}) it will be interesting to investigate the collective mode spectrum in this case for different admixture values of *s*–wave state (d_{xy}– state). Considered by Brusov et al. particular case of $d_{x^2-y^2} + id_{xy}$ state [1] shows that such consideration leads to very interesting results. One more possibility is connected with the recent experiments in Sr_2RuO_4 where the p–pairing appears to have been precluded by recent NMR experiments, the two–component d–wave order parameters, namely $\{d_{xz}, d_{yz}\}$ and even with admixture of g–wave $\{d_{x2-y2}, g_{xy(x2-y2)}\}$, are now the prime candidates for the order parameter of the quasi–two–dimensional Sr_2RuO_4. So, it will be interesting to study the collective mode spectrum in such states.

5. Conclusions

We consider all superconducting states, arising in symmetry classification of p-wave and d-wave 2D–superconductors, and calculate the full collective modes spectrum for each of these states.

The collective mode spectrum could manifest itself in microwave impedance technique, in ultrasound experiments, ultrasound velocity measurements and others. They allow determine the type of pairing and the symmetry of order parameter in HTSC and HFSC.

Author details

Peter Brusov[1]* and Tatiana Filatova[2]

1 Department of Mathematics, Financial University under the Government of Russian Federation, Moscow, Russia

2 Department of Financial and Investment Management, Financial University under the Government of Russian Federation, Moscow, Russia

*Address all correspondence to: pnb1983@yahoo.com

References

[1] Brusov Peter, Paul Brusov (2009) Collective Excitations in Unconventional Superconductors and Superfluids, 860 p. World Scientific Publishing.

[2] Ghosh, S., Shekhter, A., Jerzembeck, F. et al. Thermodynamic evidence for a two-component superconducting order parameter in Sr2RuO4. Nat. Phys. (2020). https://doi.org/10.1038/s41567-020-1032-4

[3] Benhabib, S., Lupien, C., Paul, I. et al. Ultrasound evidence for a two-component superconducting order parameter in Sr2RuO4. Nat. Phys. (2020). https://doi.org/10.1038/s41567-020-1033-3

[4] Agterberg, D.F. The symmetry of superconducting Sr2RuO4. Nat. Phys. (2020). https://doi.org/10.1038/s41567-020-1034-2

[5] Peter Brusov and Tatiana Filatova (August 24th 2015). How to Distinguish the Mixture of Two D-wave States from Pure D-wave State of HTSC, Superconductors – New Developments, Alexander Gabovich, IntechOpen, DOI: 10.5772/59180. Available from: https://www.intechopen.com/books/superconductors-new-developments/how-to-distinguish-the-mixture-of-two-d-wave-states-from-pure-d-wave-state-of-htsc

[6] Brusov P. N. (1999), Mechanisms of High Temperature Superconductivity", v.1,2; Rostov State University Publishing, p.1384.

[7] Brusov P.N., V.N.Popov (1981), Superfluidity and Bose-excitations in He3 films Sov. Phys. JETP, 53(4), 804-810.

[8] Brusov P.N., V.N.Popov (1982), Superfluidity and Bose-excitations in He3 films Phys.Lett., **87A**, #9, 472.

[9] Brusov P.N., and N.P.Brusova (1995), The model of d-pairing in CuO2 planes of HTSC and the collective modes. J. Low Temp. Phys. 101, 1003.

[10] Brusov P.N., N.P.Brusova (1994), The collective excitations in Cuo2 planes of HTSC under d-pairing, Physica C, 235-240.

[11] Brusov P.N., N.P.Brusova, P.P. Brusov, N.N.Harabaev (1997), "The path integral model of d-pairing in CuO2 planes of HTSC and the collective modes", Physica C, **282-287**, p.1833-1834.

[12] Brusov Peter, Paul Brusov and Chong Lee (2004), Collective properties of unconventional superconductors, Int. J. of Mod. Phys. B 18, 867-882.

[13] Brusov, P. N., N. P. Brusova (1994), Physica B 194-196, 1479.

[14] Brusov, P. N., N. P. Brusova, P. P. Brusov (1996), Czechoslovak Journal of Physics, 46, suppl. s2, 1041.

[15] Tewordt L. (1999), Phys. Rev. Letts. 83, 1007.

[16] Higashitani S., and K. Nagai (2000), Phys. Rev. B 62, 3042.

Chapter 7

Periodogram Analysis under the Popper-Bayes Approach

George Caminha-Maciel

Abstract

In this chapter, we discuss the use of the Lomb-Scargle periodogram, its advantages, and pitfalls on a geometrical rather than statistical point of view. It means emphasizing more on the transformation properties of the finite sampling – the available data – rather than on the ensemble properties of the assumed model statistical distributions. We also present a brief overview and criticism of recent literature on the subject and its new developments. The whole discussion is under the geophysical inverse theory point of view, the Tarantola's combination of information or the so-called Popper-Bayes approach. This approach has been very successful in dealing with large ill-conditioned, or under-determined complex problems. In the case of periodogram analysis, this approach allows us to manage more naturally the experimental data distributions and its anomalies (uncorrelated noise, sampling artifacts, windowing, aliasing, spectral leakage, among others). Finally, we discuss the **Lomb-Scargle-Tarantola (LST) periodogram**: an estimator of spectral content existing in irregularly sampled time series that implements these principles.

Keywords: Lomb-Scargle, periodogram, irregular sampling, inverse theory, spectral analysis, cyclostratigraphy, paleoclimatology, pattern recognition

1. Introduction

Although being old, periodogram analysis until nowadays represents the main workhorse for the studying of irregularly sampled time series from a vast majority of scientific branches. Since its introduction more than a century ago by A. Schuster, the periodogram has evolved and gained widespread use, even if sometimes without a complete understanding of its more subtle aspects. Its popularity comes from its relatively simple statistical behavior, easy implementation, and easy interpretation of the results. In summary, the Lomb-Scargle periodogram is an estimation method that emulates the power of Fourier decomposition – in a case when it is not possible to apply it – for data series irregularly sampled in time.

Being unnecessary to advocate for Fourier analysis, we want here to remember one of its main advantages – its *simplicity*. Fourier basis, sines and cosines, are the most basic functions that exhibit periodic behavior. Then it is very natural to use a Fourier basis to compare and detect periodic patterns in experimental data.

Usually applied in areas as diverse as astronomy, biology, meteorology, oceanography, and cyclostratigraphy [1–7], the Lomb-Scargle periodogram has not, however, a unique direct rule of use. We should always consider the subtle differences in the time series from each of these areas to explore better the spectral content in the data set and to improve understanding of the results.

There are fundamental questions that still permeate the whole subject of periodogram analysis:

- What are the necessary conditions on the original continuous function for the periodogram to analyze the irregularly sampled time series?
- What is the relationship between the Lomb-Scargle periodogram [8–14] and the Discrete Fourier Transform (DFT)?
- What is the appropriate discrete domain of frequency numbers for which an irregularly sampled time series has information? What is the minimum frequency allowable? Is there a maximum frequency (Nyquist limit) to the analysis of an unevenly sampled time series, and which it would be? What is the proper density of frequency points?
- What is the source of the several spurious peaks arising in the periodogram, besides the original peak on the proper periodicity frequencies?
- What is the uncertainty in the frequency of the periodicity found?

In this chapter, we discuss these questions and present a Popper-Bayes point of view of the periodogram, comparing it with the more traditional approach. See [15] for a comprehensive review (see also https://jakevdp.github.io/blog/2017/03/30/practical-lomb-scargle/). The traditional approach of the Lomb-Scargle method was developed mainly by astronomers and was adapted to applying to the characteristics of the astronomical data. Here we point out that the techniques devoted to astronomy are non-unique and not necessarily appropriate to other areas. We also show examples from cyclostratigraphy, our subject of study, which has some typical sampling anomalies.

Then we will present the **Lomb-Scargle-Tarantola (LST) periodogram** [16, 17], a more general technique for the use of the periodogram. The LST periodogram applies the Popper-Bayes perspective to the periodogram of irregularly sampled time series: incorporate the *a priori* variance (sampled data with all of its anomalies) directly into the *a posteriori* (periodogram) variance, and analyze the ill-defined, possibly multi-modal, complex obtained distribution.

2. The Popper-Bayes approach to inverse problems

In geophysics, it is usual to deal with high-dimensional and ill-conditioned problems. This happens because geophysicists are always trying to understand and image subsurface structures with data generally obtained from the surface. Furthermore, the measurements we want to interpret are, in general, indirectly related to the structure we want to model.

In gravimetry, for example, we get measurements of the gravitational field at a spatial grid on the surface and try to figure out what possible subsurface density anomalies could produce the observed gravitational field anomaly. This problem is highly ill-conditioned since an infinite number of configurations of subsurface bodies and density contrasts could originate the same set of field measurements at the surface. It cannot be solved without adding an *a priori* information or some kind of regularization.

To solve this class of problems, geophysicists developed statistical techniques collectively called *inverse theory*. In the inverse theory, we deal with two main difficulties: first, to find at least one model that satisfies the measurements; second, to qualify the set of obtained models. Usually, we find the **best model** approach to attack this problem – where we pick a model from a subset of possibilities by maximizing some kind of measure over the whole set of possibilities. After that, we attach some uncertainty to the best model chosen.

This approach is favored by statisticians and can be mathematically formalized. Generally, it works well when the involved variables, data and noise, follow regular statistical distributions. However, in the case of very irregularly sampled short time series, this assumption departs from reality.

2.1 Progressing by falsification

A Bayesian attack on the problem would be by posing the following question: how does this newly obtained data set modify our previous knowledge about the system?

Furthermore, it means to think about what we already know about the system – usually put in the form of probability distributions on the dynamical variables – and how to incorporate new information through the means of the constitutive equations on the dynamics. After what, we arrive on the *a posteriori* distributions (or models) "assimilating" the new data.

Here we present a more radical idea of physical inference, which is called the Popper-Bayes approach, which departs entirely from the idea of finding the best model, the mean model, or the maximum likelihood model. Professor Albert Tarantola started this idea: observations might not be used to produce models; they should be used only to falsify models [18–20].

He proposed that physical inference could be set in principle as:

1. Using the available *a priori* information to create all possible models on the system – potentially an infinite number of them;

2. For each model, solve the direct problem – assuming as true, calculate a measure (or probability) for this model in comparison with the actual observations;

3. Use some criteria to define which models are acceptable based on these measures (or probabilities) and the physical theory on the system. The unacceptable models should be dropped or falsified;

4. The set of surviving models constitute **the solution** for the physical inference. Uncertainties on these models should consider the properties of these *a posteriori* distributions over the variables subspace.

Then we have a natural interpretation of multi-modal probability distributions or ill-defined final models – what is very useful in periodogram analysis of unevenly sampled times series.

3. Periodogram analysis of irregularly sampled time series

What are the needed assumptions for properly analyzing a continuous signal by the Fourier method through a discrete sampling?

First of all, it has to be a *single function* of the time variable t (t can also be a spatial variable). It means that the signal needs to be a unique sequence of values $(x(t), t)$, where, for each t, there is one and only one assigned value $x(t)$. Besides that, the values $x(t)$ cannot "explode" (as the exponential function) – they all have to be limited by two real numbers: a maximum and a minimum. Furthermore, the function $x(t)$ cannot oscillate too fast; it has to be relatively smooth. This last condition means that the function cannot have "jumps"; in Fourier analysis, this would mean that the function has limited informational content or is "bandlimited." Our discussion here uses real-valued functions, but the same ideas can be easily extended for complex-valued cases.

The Fourier transform is a mathematical tool that for which a real function $x(t)$ relates another, a complex function – $X(f)$. As a complex number, each $X(f)$ value can be described by a pair of real numbers – an amplitude $A(f)$ and a phase $\theta(f)$. These values $A(f)$ represent the relative importance of each time scale T ($f = 1/T$). The squared values $A^2(f)$ are proportional to the *relative energy* of each of the frequencies f in the signal. The function that gives the relative energy of each of these frequency components is called *power spectral density (PSD)* $P_X(f)$, where $P_X(f) = A^2(f)$. It is usual, in the literature, to find these expressions in terms of the variable ω, called *angular frequency*, instead of f, where $\omega = 2\pi f$.

3.1 The periodogram

The Fourier transform has an analytical form applicable to continuous functions and also has a discrete form applicable to discrete functions – as the sampled time series we intend to study. That one is called *Discrete Fourier Transform (DFT)*. Being computationally intensive, particularly for large data sets, the DFT has a fast implementation called *Fast Fourier Transform (FFT)* algorithm. The FFT algorithm dramatically reduces the time and the computational cost of calculating Fourier transforms for real-time series. It is worth mentioning that if the time series is irregularly sampled, the FFT algorithm cannot be applied.

From the DFT, we can obtain estimates for the PSD of experimental real-time series. The statistics that gives an estimation of the relative energy among the different frequencies present in a signal is called the *periodogram*.

The classical periodogram is simply the squared modulus of the DFT. In the exponential form, it can be written as

$$P_X(\omega) = \frac{1}{N_0}\left|\sum_j \left(x_j \cdot e^{-i\omega t_j}\right)\right|^2 \tag{1}$$

Or in the trigonometric (equivalent) form as

$$P_X(\omega) = \frac{1}{N_0}\left[\left(\sum_j x_j \cos \omega t_j\right)^2 + \left(\sum_j x_j \sin \omega t_j\right)^2\right] \tag{2}$$

where N_0 is the number of data points in the time series.

One main statistical property of the classical periodogram is that for a time series constituted solely of evenly spaced Gaussian noise, the values of the periodogram are exponentially distributed. Unfortunately, when the time series is irregularly sampled in time, this property no longer holds. This statistical behavior also only applies to observations of uncorrelated white noise.

Lomb [9] and Scargle [10] addressed the problem of finding a generalized form of periodogram in order to:

- Reduces to the classical form in the case of uniformly sampled time series;
- Its statistics are computable;
- It is invariant to global time-shifts in the series.

The classical periodogram is very noisy even for time series only slightly noisy. The Scargle's modified periodogram, called *Lomb-Scargle periodogram*, is much smoother and differs from the classical periodogram in at least two aspects:

- It adds a time-shift term τ. This time-shift is calculated to minimize independence between the two trigonometric basis $(\sin \omega t_j)$ and $(\cos \omega t_j)$. In other words, it minimizes the crossing term $\left(\sum_j \sin \omega t_j \cos \omega t_j\right)$. It is an attempt to improve orthogonality in the equations.
- It adds the denominators $\left(\sum_j \sin^2 \omega(t_j - \tau)\right)$ and $\left(\sum_j \cos^2 \omega(t_j - \tau)\right)$ to the terms in periodogram. These denominators differ from $N_0/2$, which is the expected value in the limiting case of complete phase sampling at each frequency (as in the uniformly sampled series).

The Lomb-Scargle periodogram is given by

$$P_X(\omega) = \frac{1}{2}\left\{ \frac{\left[\sum_j x_j \cos \omega(t_j - \tau)\right]^2}{\sum_j \cos^2 \omega(t_j - \tau)} + \frac{\left[\sum_j x_j \sin \omega(t_j - \tau)\right]^2}{\sum_j \sin^2 \omega(t_j - \tau)} \right\} \tag{3}$$

where τ is given by

$$\arctan(2\omega\tau) = \frac{\left(\sum_j \sin 2\omega t_j\right)}{\left(\sum_j \cos 2\omega t_j\right)} \tag{4}$$

With this formulation, the periodogram of uncorrelated irregularly spaced Gaussian noise is also exponentially distributed (sum of squares of two zero-mean Gaussian variables).

3.1.1 The least-squares periodogram and its extensions

The Lomb-Scargle periodogram is equivalent to the least-squares fitting of a sinusoidal model to the data at each frequency ω. The Lomb-Scargle periodogram power relates to the $\chi^2(\omega)$ goodness-of-fit, at the frequency ω.

Let us consider a sinusoidal model at the frequency ω,

$$y(t;\omega) = A_\omega \sin(\omega(t - \phi_\omega)) \tag{5}$$

The χ^2 goodness-of-fit can be defined as

$$\chi^2(\omega) \equiv \sum_j \left(y_j - y(t_j;\omega)\right)^2 \tag{6}$$

We can find the "best" model $\hat{y}(t;\omega)$ by minimizing χ^2. Let $\hat{\chi}^2$ be the minimum and $\left(\hat{A}_\omega; \hat{\phi}_\omega\right)$ the optimal value, then we can write

$$P_{LS}(\omega) \sim \hat{A}_\omega^2 \tag{7}$$

For data sets with errors, we can consider introducing them into the periodogram. Vio and others [16, 17] studied a more general model, including a $N \times N$ error covariance matrix Σ, for N timely observations.

$$\chi^2(\omega) = \left(\vec{y} - \vec{y}_{model}\right)\Sigma^{-1}\left(\vec{y} - \vec{y}_{model}\right) \tag{8}$$

In the case of uncorrelated zero-mean colored noise, this expression reduces to

$$\chi^2(\omega) = \sum_j \left(\frac{y_j - y(t_j;\omega)}{\sigma_j^2}\right)^2 \tag{9}$$

where σ_j^2 are the gaussian errors.

For practical applications, there are some additional issues to consider when introducing data errors in the periodogram calculations, such as unaccountable uncertainties in error estimates, correlated noise, and the dependence in the signal slope. All of that makes the use of error estimates not very advisable [13].

3.2 Periodograms and significance

The Lomb-Scargle periodogram keeps most of the optimal analytical properties of the Fourier transform and its power spectrum:

- Linearity;
- The periodogram of a pure sinusoidal at ω_0 is a sum of the Dirac delta functions at $(\pm\omega_0)$;
- The periodogram, just as the power spectrum, is insensitive to translations in time (only reflects on the phase spectrum);
- It is a real-valued even function (that is the reason why we only calculate the positive part);
- The transform of a gaussian is another (different) gaussian;
- The "Heisenberg uncertainty principle" of Fourier transforms (usual in quantum mechanics) applies: a narrow feature in time becomes a broader peak in frequency and vice-versa.

3.2.1 Spectral windows

The pointwise product of the underlying infinite periodic signal with a rectangular window function usually describes the observed signal; its length is the time series duration T.

In the Fourier transform, the windowing replaces each Dirac delta function at some frequency ω_i with a *sinc* function centered at that same frequency ω_i, in the Fourier transform. This behavior is a direct consequence of the inverse relationship between the time window width and the width of its Fourier transform.

An infinite $\sin(\omega_i t)$ function runs from $-\infty$ to $+\infty$ and has two delta functions at $\pm\omega_i$ as its Fourier transform. The finite signal, the windowing version, has a broader transform – the *sinc* function. The *sinc* function, besides having a broader and lower *central peak* (delta function has infinite height), also has *side lobes*. It spreads the power at the frequency ω_i to the adjacent frequencies. This phenomenon is called *spectral leakage*, and it is more pronounced as the shorter is the duration of the time series (narrower window).

There is another essential aspect of spectral windows: its smoothness. The more abrupt (less smooth) is the window, the more spectral leakage happens in the Fourier transform, lowering the central peak and heightening the side lobes. Instead of a rectangular function, we can use a smoother function, like a $\sin$ bell function, for example, and the resulting spectrum will exhibit much less leakage.

Though we made this discussion about the spectrum power, the Fourier transform squared modulus, it equally applies to its actual estimate from data – the periodogram.

3.2.2 Frequencies for periodogram analysis

There are three parameters to consider when choosing the appropriate frequency grid for periodogram analysis of a particular time series:

1. The minimum frequency, f_{min};

2. The maximum frequency, f_{max};

3. The frequency spacing, Δf.

Recommendations for the choice of each of these parameters vary in the literature. Here we discuss some points to consider in the case of regularly as well as to irregularly sampled time series.

Minimum frequency: The minimum frequency, f_{min}, is the easiest to define. It relates to the largest period of a wave we can investigate in the time series. We usually set it as the inverse of T – the length of the time series, or as zero – where its value virtually equals to the frequency spacing, Δf.

Maximum frequency: The maximum frequency, f_{max}, represents the shortest period of a wave we can investigate in the time series. For evenly sampled time series, the *Nyquist theorem or Sampling theorem* defines this maximum frequency – called *Nyquist frequency*, $f_{Nyquist}$.

This theorem states that if we have a regularly sampled function with the sampling rate of $f_\delta = 1/\delta t$, we can only recover full frequency information if the signal is *band-limited* between frequencies $f_\delta/2$. This theorem states that if we have a regularly sampled function with the sampling rate of $f_\delta = 1/\delta t$, we can only recover full frequency information if the signal is *band-limited* between frequencies $\pm f_\delta/2$.

Putting in another way, the theorem says that to fully represent the content of a band-limited signal whose Fourier transform is zero outside the range of $\pm B$, we must sample the signal with a rate at least $f_\delta = 2B$. Then, for evenly sampled time series, $f_{max} = f_{Nyquist} = f_\delta/2$.

Frequency spacing: The frequency spacing, Δf, has only general guidelines: too small frequency spacing can lead to unnecessarily long computation times, which adds up fastly for large data sets. Too coarse frequency spacing can risk missing narrow peaks in the periodogram – which would fall between adjacent grid points. However, there is a controversy when considering these frequency grids as independent points when applying statistical significance tests in the periodogram ordinates (testing for true periodicities).

An evenly sampled time series represents a pointwise product of the original continuous signal with a sequence of Dirac delta functions (a *Dirac comb*) at the sampling times. The Nyquist limit is a direct consequence of the symmetry in this Dirac comb window. Beyond this limit, the spectrum becomes a periodic repetition of itself – that is why the periodogram is unique between the limits $\pm f_{Nyquist}$. The rise of power in the spectrum beyond the Nyquist limit is called *aliasing* since these peaks are not real but "alias" of the real power inside the Nyquist interval in the original signal.

For unevenly sampled time series:

- The structure in the observing window can lead to *partial aliasing* in the periodogram;
- The non-structured spacing of observations also leads to the arising of *non-structured peaks* in the window transform;
- The maximum frequency limit *might or might not exist*, and if it exists, it tends to be *far higher* than for the evenly sampled case.

For irregularly sampled time series, if there is a *periodic pattern in the observation times gaps*, this can lead to a peak in the periodogram indicating a periodicity. For example, the daily pattern of measurements in astronomy: an observation in time t_0 is likely to be followed by other observation only at time $t_0 + np$ (p is an integer number of days, and n is an integer). Therefore, it can generate a peak at the frequency p in the periodogram.

We find in the literature some proposals for the maximum frequency (Nyquist-like) limit for irregular sampling [15, 21–26]. These estimates are easy to calculate and reduce to the Nyquist frequency limit in the evenly sampled case:

- Arithmetic average of the time intervals.
- Harmonic average of the time intervals.
- Median of the time intervals.
- Minimum of the time intervals.

There are also, in the literature, some Nyquist-like limits based on not-so-simple statistics of the time intervals [15]:

- Greatest common divisor (gcd) of the time intervals. We should consider sampling times as integer numbers (what can always be done by re-scaling the time values).
- Frequency limit due to *time windowing* when observations are not pointwise instantaneous, but instead, they consist of *short-time* (δt) *integrations* of the

continuous signals, around the sampling times t_j. This kind of sampling is typical in several applications, including cyclostratigraphy. In that case, again, the Fourier transform is the product of the original signal transform and the transform of the time window, which has the width proportional to $1/\delta t$. Then, $f_{max} = f_{Nyquist} = 1/(2\delta t)$. However, in this case, this frequency limit does not imply aliasing. Instead, it is about a frequency limit beyond which all signal is *attenuated to zero*.

- Frequency limit based on *a priori* knowledge on the expected signals.

Finally, for irregularly sampled time series, the maximum frequency limit can be set by the *precision of time measurements*.

3.2.3 Statistics of the periodogram

The classical periodogram has a fundamental statistical property for evenly sampled time series: when the signal consists solely of pure Gaussian noise, the values of the periodogram are exponentially distributed – for irregularly sampled times, this property no longer holds.

The Scargle's generalized form of the periodogram brings back that statistical simplicity for the irregularly sampled case: for time series consisting solely of pure Gaussian noise, the unnormalized periodogram has its ordinates exponentially distributed.

This statistical property is used to test for what would be a "true" periodicity in periodogram ordinates. The standard procedure is to assume that the periodogram maximum ordinate represents a true periodicity, called *Fisher criteria*, and to test this value against all others ordinates – supposedly arising from the background noise.

Scargle defined a *False Alarm Probability (FAP)* that, based on the assumed distribution of Gaussian noise, simply measures the probability that a time series without any signal would arise, due to stochastic fluctuations only, an ordinate of the observed magnitude in the periodogram. Following Scargle [15], the *detection threshold*, z_0, is a magnitude level above which, if we claim that a peak is due a real signal, we would only be wrong a small fraction p_0 (FAP) of the time:

$$z_0 = -\ln\left[1 - \left(1 - p_0\right)^{1/N}\right], \tag{10}$$

where p_0 (FAP) is a small number, and N is the number of *independent* frequencies tested.

It is worth noting that this statistical analysis answers the question: "What is the probability that a time series without any periodic component would make arise a peak of that magnitude in the periodogram?" It *does not* answer the utterly more physically significant, more direct question: "*What is the probability that this periodogram feature comes from a periodic phenomenon?*"

The ability to analytically quantify the relationship between peak height and statistical significance of a feature in the periodogram has been one of the main reasons for the widespread use of the Lomb-Scargle periodogram [10–13, 23–25]. However, the independence of the tested frequencies remains an open issue.

Data quality (and quantity) generally reflects on the peak height related to the background noise, which gives peak *significance*, as discussed above. Neither the number of points in a time series or the signal-to-noise ratio affects the peak frequency determination nor its *precision*. The uncertainty in the frequency value of a peak is related to the peak width, usually in Fourier analysis defined as the peak half-width. For this reason, in periodogram analysis, Gaussian error bars should be avoided as a way to report uncertainties in frequency determinations.

4. The Lomb-Scargle-Tarantola (LST) periodogram

In geophysical inverse theory, there are high mature methods to deal with large uncertainties in data and ill-conditioned models. It comes from the constant necessity; geophysicists have to build and evaluate physical models of the subsurface (or Earth's deep interior) based mostly on data acquired from the surface.

In Seismics, for example, geophysicists developed a procedure called *stacking* where a set of different signals, acquired by distinct geophones, are gathered under specific geometrical settings. The procedure aims to amplify the signal and to cancel noise, improving overall information about the same subsurface points. The stacking of seismic signals is one of the main reasons for the success of this technique on the oil industry nowadays, allowing the discovery of new oil fields and improving oil/gas recovering on reservoir structures.

Caminha-Maciel and Ernesto [16, 17] created a method to analyze spectral content (and its uncertainties) in irregularly sampled times series applying some principles from the geophysical inverse theory – the Popper-Bayes approach, as developed by Albert Tarantola [18–20]. This approach uses freely normalizable probability distributions to encapsulate the information and afterward operates with these distributions (the data information and the model/geometrical information on the problem). The results of these operations constitute a Bayesian physical inference method, and its *a posteriori* probability distributions are proper solutions to our inverse problem. In the following section, we will see how this applies to the Lomb-Scargle periodogram.

4.1 State of information periodogram-based functions

There is a vast diversity of methods to analyze time series, both in time and in frequency domains. Fourier-derived methods still show continued interest since they are fast (due to FFT algorithm), intuitive, and have a straightforward extension to irregularly sampled time series.

However, in some applications, as the high-resolution deep-sea stratigraphic records, there are novel challenges for the extraction and interpretation of meaningful information. It is worth to mention the uncertainties in the measured times (plus the usual uncertainties on the other variables), non-stationarity of the dynamical systems observed, and shortness of the records compared to wavelengths of interest. In the stratigraphic time series, there is also a general dependence of the recording-sampling process on the unknown-climatic signal itself. These issues contribute to breaking the orthogonality among periodogram ordinates, which is necessary to perform appropriately statistical significance tests on supposed-independent ordinates.

Nevertheless, there are some less-known interesting properties of the Lomb-Scargle periodogram:

- Analytical independence among ordinates – non-statistical independence. Since we have a fixed set of points $(X(t_j), t_j)$ and a fixed frequency point ω_0, $P(\omega_0)$ gives the same result regardless of how many different frequencies ω_i we use in the analysis. The periodogram does not "see" other ordinates. The problem is when we try to compare different ordinates (statistical independence).

- The maximum frequency (Nyquist-like) allowed can be much higher than in the evenly sampled case. We can define a Nyquist-like frequency related to the inverse of the minimum time interval, or even the gcd of the time intervals [8].

- The signal-noise (S/N) ratio does not depend on the periodogram normalization factor. In other words, $P(\omega_a)/P(\omega_b)$ is independent of the chosen normalization factor (there are several different options in the literature).

- There is a low false-negative probability for the periodogram for a real signal present in the time series. Then, if we guess correctly a frequency ω_0 present in the time series, we have a very low probability of obtaining a small value for $P(\omega_0)$. The problem is that we also obtain high periodogram ordinates for other frequencies besides the true ones – this makes it difficult to identify them if we do not know beforehand.

In the LST periodogram, a freely normalized version of the Lomb-Scargle periodogram is initially defined over the broadest possible range of frequencies. The frequency set is chosen to include all wavelengths about which could exist information on the time series. The next step is to choose the minimum frequency as zero, the frequency grid spacing δf, or the total length of time series, or some fraction in between these values. We can choose the maximum frequency as the highest frequency about which we believe there is any information in the time series, as some *a priori* known Nyquist limit, up to the limit of the inverse of the gcd of time intervals. Above this limit, we probably have to deal we some folding in the periodogram. The frequency grid δf has to be chosen to allow the calculation to be computationally feasible.

4.1.1 Normalizing by the bandwidth total content

We define

$$P_{LST}(\omega) = K \cdot P_{LS}(\omega) \tag{11}$$

where the normalizing constant K is set as $K = \left(\sum_\omega P(\omega)\Delta\omega\right)^{-1}$. This values of the constant K is such that the function P_{LST} normalizes to total area under the curve equals to 1 over the whole set of frequencies ω_i. This area represents the total power in the bandwidth and reflect the times series total variance.

Note that the S/N ratio, as well as the ratio between any two distinct frequencies $P(\omega_a)/P(\omega_b)$, does not depend on the times series total power.

This procedure is equivalent to a stretching of the data series variable $X(t)$ in the time domain and also has the property of making comparable the total power of the various periodograms.

4.2 Periodogram analysis by combination of information

The two main ideas of the Lomb-Scargle-Tarantola periodogram are

1. Smoothing the periodogram.

2. Stacking independent periodogram estimates.

Since its proposition, the periodogram is recognized as high noisy statistics, even for less noisy data. Smoothing the periodogram is not a new idea. There are several attempts in this direction by averaging adjacent estimates (as in Daniel's averaged periodogram, for example) and in many Bayesian formulations.

A new idea on working with periodograms is stacking in the frequency domain: we can consider two or more distinct time series with information about the same dynamical variable as independent observations of the same phenomenon. For example, several stratigraphic sections covering the same time interval; different variables related to the same dynamics – sediment accumulation and $\delta^{18}O$ time series, both related to seawater surface temperature; independent observations of the same variable – as astronomical observations from distinct geographic locations; and biological circadian rhythms from different organisms.

After that, we operate these periodogram distributions with two logical operators "OR" and "AND." The OR operator can be described as a generalization of doing histograms and is mathematically defined as the arithmetic average of the individual LST periodograms (very similar to usual stacking of seismic signals). The AND operator is a non-linear operator that represents the generalization of conditional probability and is mathematically defined as $\frac{\Pi_i P_i(\omega)}{\mu}$, where μ is the null information function – characterizing the geometry of the physical problem.

In some physical problems involving the dynamical variable *frequency*, f, itself, the null information function can be better written as $1/f$. However, in spectral analysis, the variable frequency only means labels for some general class of eigenfunctions – as the Fourier basis sin and cos. Then we can fairly consider the null information function, μ, as a *constant function* over the entire domain.

4.2.1 Using LSTperiod software

We have published [17] a proof-of-concept software to implement the LST periodogram. This software, the *LSTperiod* (Download it at http://www.iag.usp.br/paleo/sites/default/files/LSTperiod-files.zip), exemplify one possible implementation of the LST periodogram. We have made a set of choices for the frequency grid, normalizations of the state of information functions, evaluation of the resulting models (candidates periodicities), and visualizations of the results. Those choices are for no means unique [27].

To illustrate the use of the LSTperiod software, we show a set of five stratigraphic series of benthic $\delta^{18}O$ from the sedimentary core drilled by the Ocean

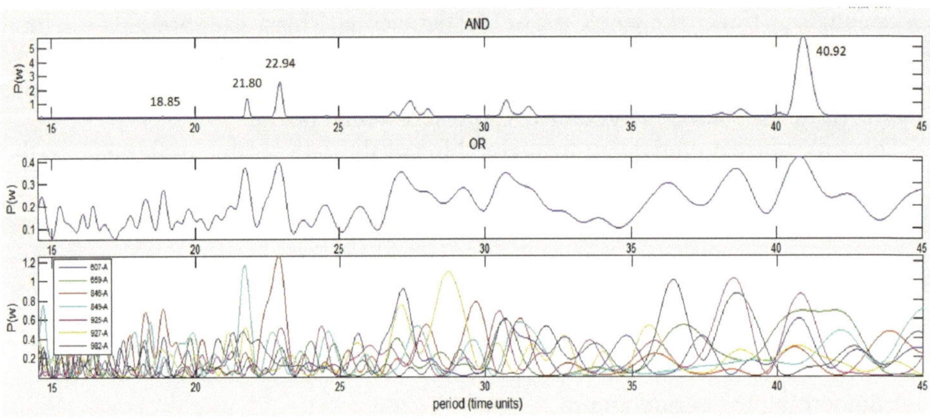

Figure 1.
LST periodograms for the benthic $\delta^{18}O$ series from ocean drilling project (ODP) cores. The windows show the periodogram calculated for each data file separately (bottom) and the combined results for the OR (middle) and AND (top) spectra.

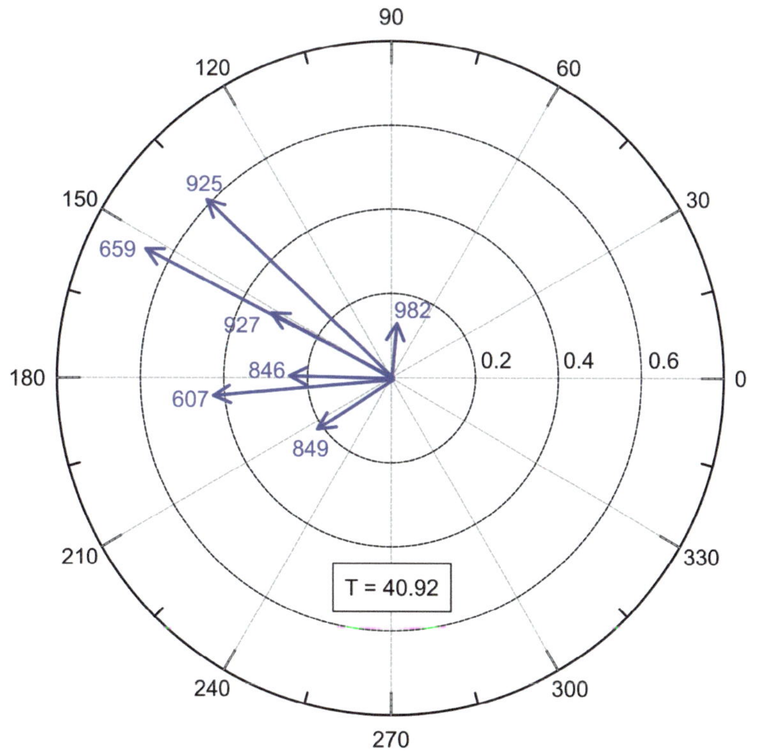

Figure 2.
Rose diagram showing the phases and amplitudes of the calculated period $T = 40.92kyr$ *for each analyzed series.*

Drilling Project (ODP) [28]. These time series were subjected to spectral analysis and other statistical methods and show Milankovitch climatic cycles around 19, 23, and 41 kyr [28–30].

In **Figure 1**, we can see the periodograms for these time series – individual and combined (OR/AND) periodograms. **Figure 2** shows the amplitude and phase analysis for the $\sim 41kyr$ Milankovitch period found, for each analyzed series.

5. Conclusions

With the recent advances in experimental sciences, there is an increased need to analyze and statistically evaluate the information contained in time series. In some areas, as paleoclimatic studies, the study of the information on this kind of data constitutes the main body of physical evidence to understand and solve fundamental today's problems – as the origin and development of the *climatic change* observed in recent years.

Fourier methods still constitute an updated tool since they are simple and offer easy-to-understand results. However, in several of these applications, as in stratigraphic data, these time series come with severe sampling anomalies. These anomalies ultimately prevent the use of standard Fourier techniques, as the FFT algorithm and periodograms. The Lomb-Scargle periodogram has been very useful to attack this sort of problem. However, its use, as seen in the literature, lacks a proper analysis of the uncertainties associated and, worse, is unfit to be applied in the most poorly sampled time series.

The spectral analysis of irregularly sampled times series represents a problem of studying a physical system from incomplete information. As we know, from geophysical inverse theory, this kind of problem cannot be solved without some input of *a priori* information – explicitly or not.

The Popper-Bayes approach for physical inference proposes to solve inference problems by a combination of information: theoretical usually expressed through the functional form of statistical distributions and by combining independent experimental data. The LST periodogram is a development of the Lomb-Scargle periodogram that brings these inference principles to the periodogram analysis of irregularly sampled time series. The main idea of the LST periodogram is to smooth the periodogram of a dynamical variable through the stacking of spectral information from multiple irregularly sampled times series.

The periodogram of an irregularly sampled times series cannot, by any means, become a set of independent ordinates for being submitted to a proper statistical test. With the LST periodogram, we propose to change the use of the periodogram: from an auxiliary tool to statistical decision theory (define a periodicity) to a dimension reduction problem – from a broad set of possible frequencies to very narrow set of periodogram local maxima (peaks).

Acknowledgements

We are thankful to Dr. M. Ernesto for critically reading the early version of this manuscript. Thanks are also due to the editor Dr. Y.K.N. Truong whose suggestions greatly improved the manuscript.

Note

The Lomb-Scargle-Tarantola (LST) periodogram was very much inspired by the seminal work of Professor Albert Tarantola (1949–2009) on Physical Inference and Geophysical Inverse Problems.

Author details

George Caminha-Maciel
Computational Geophysics Laboratory (COMPUTAGEO), Geology Department, Universidade Federal de Santa Catarina (UFSC), Florianópolis, SC, Brazil

*Address all correspondence to: caminha.maciel@ufsc.br

References

[1] Péron G, Fleming CH, de Paula RC, Calabrese J. Uncovering periodic patterns of space use in animal tracking data with periodograms, including a new algorithm for the Lomb-Scargle periodogram and improved randomization tests. Movement Ecology. 2016;**4**:19

[2] Baldysz Z, Nykiel G, Araszkiewicz A, Figurski M, Szafranek K. Comparison of GPS tropospheric delays derived from two consecutive EPN reprocessing campaigns from the point of view of climate monitoring. Atmospheric Measurement Techniques. 2016;**9**: 4861-4877

[3] Berger WH. On the Milankovitch sensitivity of the quaternary deep-sea record. Climate of the Past. 2013;**9**: 2003-2011

[4] Bowdalo DR, Evans MJ, Sofen ED. Spectral analysis of atmospheric composition: Application to surface ozone model–measurement comparisons. Atmospheric Chemistry and Physics. 2016;**16**:8295-8308

[5] Dawidowicz K, Krzan G. Analysis of PCC model dependent periodic signals in GLONASS position time series using Lomb-Scargle periodogram. Acta Geodynamics et Geomaterialia. 2016; **13**(3):299-314

[6] Nielsen T et al. Are there multiple scaling regimes in Holocene temperature records? Earth System Dynamics. 2016;**7**:419-439

[7] Hinnov LA. Cyclostratigraphy and its revolutionizing applications in the earth and planetary sciences. GSA Bulletin. 2013;**125**:1703-1734

[8] Deeming TJ. Fourier analysis with unequally spaced data. Astrophysics and Space Science. 1975;**36**:137-158

[9] Lomb NR. Least-squares frequency analysis of unequally spaced data. Astrophysics and Space Science. 1976; **39**:447-462

[10] Scargle JD. Studies in astronomical time series analysis, II, statistical aspects of spectral analysis of unevenly spaced data. The Astrophysical Journal. 1982; **263**:835-853

[11] Vio R, Andreani P, Biggs A. Unevenly-sampled signals: A general formalism for the Lomb-Scargle periodogram. Astronomy and Astrophysics. 2010;**519**:A85

[12] Vio R, Diaz-Trigo M, Andreani P. Irregular time series in astronomy and the use of the Lomb-Scargle periodogram. Astronomy and Computing. 2013;**1**:5-16

[13] Hernandez G. Time series, periodograms, and significance. Journal of Geophysical Research. 1999;**104**(10): 368

[14] Stoica P, Li J, He H. Spectral analysis of non-uniformly sampled data: A new approach versus the periodogram. IEEE Transactions on Signal Processing. 2009;**57**(3):843-858

[15] Vander Plas JT. Understanding the Lomb-Scargle periodogram. The Astrophysical Journal Supplement Series. 2018;**236**:16

[16] Caminha-Maciel G, Ernesto M. Characteristic wavelengths in VGP trajectories from magnetostratigraphic data of the early cretaceous Serra Geral lava piles, southern Brazil. In: Jovane L, Herrero-Bervera E, Hinnov L, Housen BA, editors. Magnetic Methods and the Timing of Geological Processes. London: The Geological Society of London. Special Publications; 2013. p. 373

[17] Caminha-Maciel G, Ernesto M. LSTperiod software: Spectral analysis of multiple irregularly sampled time series. Annals of Geophysics. 2019;**62**:5: DM566. DOI: 10.4401/ag-7923

[18] Tarantola A. Popper, Bayes and the inverse problem. Nature. 2006;**2**: 492-494

[19] Tarantola A, Mosegaard K. Athematical basis for physical inference. Cornell University Library. arXiv:math-ph/0009029v1; 2000

[20] Tarantola A, Valette B. Inverse problems = quest for information. Journal of Geophysics. 1982;**50**:159-170

[21] Mortier A, Faria JP, Correia CM, Santerne A, Santos NC. BGLS: A Bayesian formalism for the generalized Lomb-Scargle periodogram. Astronomy and Astrophysics. 2015;**573**:A101

[22] Mortier A, Cameron AC. Stacked Bayesian general Lomb-Scargle periodogram: Identifying stellar activity signals. Astronomy and Astrophysics. 2017;**601**:A110

[23] Munteanu C, Negrea C, Echim M, Mursula K. Effect of data gaps: Comparison of different spectral analysis methods. Annales de Geophysique. 2016;**34**:437-449

[24] Pardo-Igúzquiza E, Rodríguez-Tovar FJ. Implemented Lomb-Scargle periodogram: A valuable tool for improving cyclostratigraphic research on unevenly sampled deep-sea stratigraphic sequences. Geo-Marine Letters. 2011;**31**:537-545

[25] Pardo-Igúzquiza E, Rodríguez-Tovar FJ. Spectral and cross-spectral analysis of uneven time series with the smoothed Lomb-Scargle periodogram and Monte Carlo evaluation of statistical significance. Computers and Geosciences. 2012;**49**:207-216

[26] Zechmeister M, Kürster M. The generalised Lomb-Scargle periodogram. A new formalism for the floating-mean and Keplerian periodograms. Astronomy & Astrophysics. arXiv: 0901.2573v1 [astro-ph.IM]; 2009

[27] Townsend RHD. Fast calculation of the Lomb-Scargle periodogram using graphic processing units. The Astrophysical Journal Supplement Series. 2010;**191**:247-253

[28] Lisiecki LE, Raymo ME. A Pliocene-Pleistocene stack of 57 globally distributed benthic delta-18-O records. Paleoceanography. 2005;**20**:PA1003

[29] Jalón-Rojas I, Schmidt S, Sottolichio A. Evaluation of spectral methods for high-frequency multi annual time series in coastal transitional waters: Advantages of combined analyses. Limnology and Oceanography: Methods. 2016;**14**:381-396

[30] Lisiecki LE, Raymo ME. Plio–Pleistocene climate evolution: Trends and transitions in glacial cycle dynamics. Quaternary Science Reviews. 2007;**26**:56-69

Chapter 8

Spread Option Pricing on Single-Core and Parallel Computing Architectures

Shiam Kannan and Mesias Alfeus

Abstract

This paper introduces parallel computation for spread options using two-dimensional Fourier transform. Spread options are multi-asset options whose payoffs depend on the difference of two underlying financial securities. Pricing these securities, however, cannot be done using closed-form methods; as such, we propose an algorithm which employs the fast Fourier Transform (FFT) method to numerically solve spread option prices in a reasonable amount of short time while preserving the pricing accuracy. Our results indicate a significant increase in computational performance when the algorithm is performed on multiple CPU cores and GPU. Moreover, the literature on spread option pricing using FFT methods documents that the pricing accuracy increases with FFT grid size while the computational speed has opposite effect. By using the multi-core/GPU implementation, the trade-off between pricing accuracy and speed is taken into account effectively.

Keywords: spread option, single core, parallel computing

1. Introduction

Spread options have widespread uses across many industries, remarkably in the seasonal commodity market futures. One notable example is the crack spread, which is the pricing difference between a barrel of crude oil and the petroleum products refined from it. Traditionally, crack spreads involve the purchase of oil futures and the simultaneous sale of futures of the refined product, whether it be gasoline, heating oil, or other similar products. Refiners seek a positive spread between the prices of crude oil and refined products, meaning that the price of the input (oil in this case) is lower than the price of the output (gasoline, kerosene, etc.).

Beyond the oil industry, spread options are utilized by suppliers in industries as disparate as the soybean and electricity markets. Soybean spread options are known as crush spreads, and electricity spread options are known as spark spreads. Similar to crack spreads, both these options seek to maximize the spread between input costs and output prices for suppliers to maximize profit.

The use of spread options across such disparate fields is but a testament to their widespread use. Considering the popularity of spread options, it thus prompts the needs of accurate and fast pricing of price of these options. This paper dwells on the discussion of fast algorithms for these options.

IntechOpen

As mentioned previously, pricing spread options tends to be something that cannot be easily performed using traditional closed-form solutions. Therefore, we explore the utilization of the fast Fourier transform method to price these securities. Specifically, we perform the FFT on the spread options payoff, assuming knowledge of the model joint characteristic function, which we represent as a pointwise multiplication of the characteristic function and the complex gamma function in the Fourier domain.

Regarding our implementation, we adapt the parallel computing Toolbox in MATLAB to take advantage of the multi-core capabilities of GPU processing, to substantially improve the performance and computational efficiency of the algorithm for spread options. This methodology may serve a great deal especially in model calibration and risk management approach. For measuring performance, we only time the execution of the inverse Fast Fourier Transform, as the prior steps are merely initialization of the necessary arrays. For measuring accuracy, we compute the Euclidean norms of each of the resulting option price arrays from each implementation (single-core and GPU), and find the percent error between the norms as follows:

$$\frac{\text{GPU norm} - \text{single core norm}}{\text{single core norm}} \times 100 \tag{1}$$

The main contribution of this paper is the use of parallel computation for the spread option value using two-dimensional Fourier transform. Our implementation is developed both for single-core processors, as well as for parallel processing on multi-core/GPU systems. For both execution methods, we have implemented our algorithm using MATLAB built-in functions to produce a version of the algorithm compatible for multi-core systems. To ascertain the impact of the different environments as well as the different methods of execution on the computational efficiency of the algorithm, we record the times of execution for different values of FFT grid size N, which is the number of grid points of discretization of the characteristic function along the two asset dimensions, for both the classical single-core implementation and the multi-core/GPU implementation. This approach completely eliminates the trade off between computational accuracy and speed, that is, we price spread option accurately and in a fastest possible way.

2. Model description

2.1 Spread option valuation

We fix the trading time horizon T and consider a filtered probability space $(\Omega, \mathcal{F}, P, (\mathcal{F}_t)_{t<T<T})$ defined in the usual way.

The goal is to compute the value of an European spread option between stock price processes $S_1 = \{S_1(t)\}_{0\leq t\leq T}$ and $S_2 = \{S_2(t)\}_{0\leq t\leq T}$ with maturity time T and exercise price K.

At expiration time T, the payoff is given by

$$\mathcal{S}_T(S_1, S_2, K) = \max\{S_1(T) - S_2(T) - K, 0\}, \tag{2}$$

and under a risk-neutral conditional measure[1] Q its value at time t is given by

[1] 3 Specifically: Under the risk–neutral measure associated with taking the continuously compounded savings account as the numeraire, and (for expositional simplicity) assuming a constant interest rate r.

$$\mathcal{S}_t(S_1, S_2, K) = e^{-r(T-t)} E^Q\left[(S_1(T) - S_2(T) - K)^+ | \mathcal{F}_t\right]. \tag{3}$$

Under normality assumption, Eq. (3) can be analytically approximated as done in [1]. However, a departure from normality assumption ushers into numerical computational difficulties. In option pricing literature of finance, Fourier transform-based method is usually the best candidate to approximate the solution for Eq. (3), in this case whenever the joint characteristic function of the asset price processes, S_1 and S_2 are available. For spread options pricing valuation methodology using two-dimensional fast Fourier transform (FFT) techniques was coined in [2, 3]. We cite the important formula from [3] that gives the price for spread option

$$\begin{aligned}\mathcal{S}_t(S_1, S_2, K) &= e^{-r(T-t)} E^Q\left[(S_1(T) - S_2(T) - K)^+ | \mathcal{F}_t\right] \\ &= \frac{1}{(2\pi)^2} e^{-rT} \iint_{R+i\epsilon} e^{iuX'_0} \Phi(u, T) \hat{P}(u) d^2u\end{aligned} \tag{4}$$

$$\hat{P}(u) = \frac{\Gamma(i(u_1 + u_2) - 1)\Gamma(-iu_2)}{\Gamma(iu_1 + 1)} \tag{5}$$

$$\Gamma(z) = \int_0^\infty e^{-t} t^{z-1} dt, \tag{6}$$

where $X_0 = \log S$, $\Phi(u, T) = E\left[e^{iu(X'_T - X'_0)}\right]$, is the joint characteristic function of the log return.

Fast and accurate pricing is often the most desirable feature of the model. In [4], the authors consider spread options pricing in C++ using fast Fourier transform in the west (FFTW). They observed the trade off between fast computation and numerical accuracy; pricing accuracy is monotonic in the number of FFT grid size used in the price computation. However, using a large number of FFT grid size slow down the speed of price computation.

2.2 Model characteristic function

Fast Fourier transform (FFT) method is generically applicable in finance because it only requires the specification of the characteristic function of the random variable. In terms of spread options, one just need a characteristic function of the joint distribution of the financial variables in question. Here, we employ two characteristic functions: one based on two-dimensional normal distribution and the other one based on two-dimensional normal inverse Gaussian (NIG) distribution.

2.2.1 Two-dimensional geometric Brownian motion (GBM)

The characteristic function for a spread option comprised of two assets, each of which is modeled as a correlated GBM, is given by

$$\Phi_{\text{GBM}}(u; T) = \exp\left(iu\left(rTe - \frac{\sigma^2 T}{2}\right)' - \frac{u\Sigma_{GBM} u' T}{2}\right) \tag{7}$$

where $e = [1, 1]$, $\Sigma_{GBM} = \begin{pmatrix} \sigma_1^2 & \sigma_1\sigma_2\rho \\ \sigma_1\sigma_2\rho & \sigma_2^2 \end{pmatrix}$, and $\sigma^2 = \text{diag}(\Sigma_{GBM})$, $i = \sqrt{-1}$,. $r, \sigma_i : i = 1, 2$ denote the risk-free rate, volatilities, respectively, and ρ is the correlation parameter between two asset prices processes S_1 and S_2. Here, Σ_{GBM} is the covariance matrix.

2.2.2 Two-dimensional normal inverse Gaussian (NIG) Levy process

Let S denote a two-dimension NIG random variable. The characteristic function of $S = (S_1, S_2)$ is given by

$$\Phi_{\mathrm{NIG}}(u;T) = \exp\left(\mathrm{i}u'\mu T + \delta T\left[\sqrt{\alpha^2 - \beta\Delta\beta'} - \sqrt{\alpha^2 - (\beta + \mathrm{i}u)\Delta(\beta + \mathrm{i}u)'}\right]\right), \quad (8)$$

where $\alpha, \delta \in R_+$, $\beta, \mu \in R^2$, and $\Delta \in R^{2\times 2}$ is a symmetric, positive-definite matrix. Moreover, the structural matrix Δ is assumed to have determinant $\mathrm{Det}(\Delta) = 1$.

The covariance matrix corresponding to the two-dimensional NIG-distributed random variable S is

$$\Sigma_{NIG} = \delta\left(\alpha^2 - \beta\Delta\beta'\right)^{-\frac{1}{2}}\left(\Delta + \left(\alpha^2 - \beta\Delta\beta'\right)^{-1}\Delta\beta\beta'\Delta\right) \quad (9)$$

2.3 FFT algorithm

The FFT algorithm for spread option pricing along the line of [3] can be described as follows

Algorithm 1: FFT algorithm for Spread Option pricing

1 2-D FFT $(N, \overline{u}, \epsilon)$;
Input : N, a power of two; $\overline{u}$, truncation width; ϵ, damping factor.
Define : $u(k) = (u_1(k_1), u_2(k_2))$ and $x(l) = (x_1(l_1), x_2(l_2))$
2 Set $X_0 = \left[\log\left(\frac{S_1}{K}\right), \log\left(\frac{S_2}{K}\right)\right] \in x(l)$
3 **forall** $k, l \in \{1, \cdots, N-1\}^2$ **do**
4 $\quad H(k) = (-1)^{k_1+k_2}\Phi(u(k) + i\epsilon)\hat{P}(u(k) + i\epsilon)$;
5 $\quad C(l) = (-1)^{l_1+l_2}e^{-rT}\left(\frac{\eta N}{2\pi}\right)^2 e^{-\epsilon x(l)'}$;
6 **end**
7 $V = \mathcal{R}_e(C \times \mathrm{ifft2}(H))$ /* with $O\left(N^2 \log N\right)$ complexity*/
9 $P \leftarrow K \times V$ /* using an efficient interpolation technique*/;
Output: P

3. Numerical results

3.1 Implementation outlook

As mentioned earlier, two versions of the algorithm were programmed in MATLAB, namely a single-core variant and a multi-core GPU variant. In MATLAB, the Parallel Processing Toolbox was used to exploit multi-core GPU capabilities to run the algorithm. Among its capabilities is the ability to run for loops and perform array operations in parallel, both on multi-core CPUs as well as GPUs.

As in the Algorithm 1 given above, the for loop in lines 3–6 was run in parallel, across six CPU cores, employing the parfor directive available in the MATLAB Parallel Processing Toolbox. We also sought to run computationally heavy functions on the GPU we had available, to improve the efficiency of our algorithm beyond what would be possible on a multi-core CPU. In that regard, we executed the inverse FFT, as described in line 7, on the GPU. We accomplished this by copying our H and A arrays onto the GPU, such that any further processing of those arrays would only occur on the GPU. To perform this operation, we utilized the MATLAB inbuilt function gpuArray() and copied the two aforementioned arrays to the GPU after the for loop. To transfer the GPU results (following execution of the inverse FFT) back to the local workspace, we used the MATLAB function gather().

Table 1 shows the market parameters, and these inputs are taken from [5] where d_1 and d_2 represent the dividend rate for S_1 and S_2, respectively.

The computer used to produce the following results was an ASUS ROG Strix Scar II GL704GW, with an Intel Core i7-8750H processor clocked at 2.20 Hz and comprising of 6 cores, 16GB RAM, and an NVIDIA GeForce RTX 2070 GPU with 8GB memory, running Windows 10. The computational times of the algorithm are tabulated in **Table 2** and **Figure 1**. In a single run, we compute the price of 10 options, that is, $N_O = 10$. The pricing accuracy is gauged using the root mean square error (rmse):

$$\text{rmse} = \sum_{j=1}^{N_O} \left(\frac{P^j_{FFT} - P^j_{Monte}}{P^j_{Monte}} \right)^2, \tag{10}$$

where P_{Monte} represents the benchmark price computed using Monte Carlo method with 1000000 simulations and 1000 time step and P_{FFT} is the price from two-dimensional FFT.

S_1	S_2	r	T	d_1	d_2
100	96	0.1	1.0	0.05	0.05

Table 1.
Market pricing parameters.

Grid points: 2^N	Pricing accuracy	Single core (s)	GPU parallel (s)	GPU speed factor
(a) 2d-GBM model: $\sigma_1 = 0.2, \sigma_2 = 0.1, \rho = 0.5$				
8	1.10E-04	2.671126	0.213765	12.49561902
9	4.29E-05	10.155822	0.254708	39.87241076
10	1.97E-05	39.884087	0.650358	61.32635718
11	1.06E-05	157.833762	2.422815	65.14478489
12	6.19E 06	317.858516	9.474695	33.54815284
13	3.80E-06	1316.938937	38.519949	34.18849119
(b) 2d-NIG model: $\mu_1 = 0, \mu_2 = 0, \alpha = 6.20, \delta = 0.150\ \beta_1 = -3.80, \beta_2 = -2.50, \rho = 0, \mu_1 = 0, \mu_2 = 0$ and $\Delta = I$				
8	1.20E-04	2.528521	0.21573	11.72087536157
9	5.21E-05	9.768831	0.31099	31.41244621944
10	3.00E-05	39.372496	1.03477	38.04969824066
11	2.12E-05	162.069306	4.57217	35.44691939427
12	1.72E-05	651.850506	19.92393	32.71695934733
13	1.50E-05	2592.307476	83.77870	30.94232156861

Table 2.
Computational accuracy and processing times (a) 2d-GBM, (b) 2d-NIG.

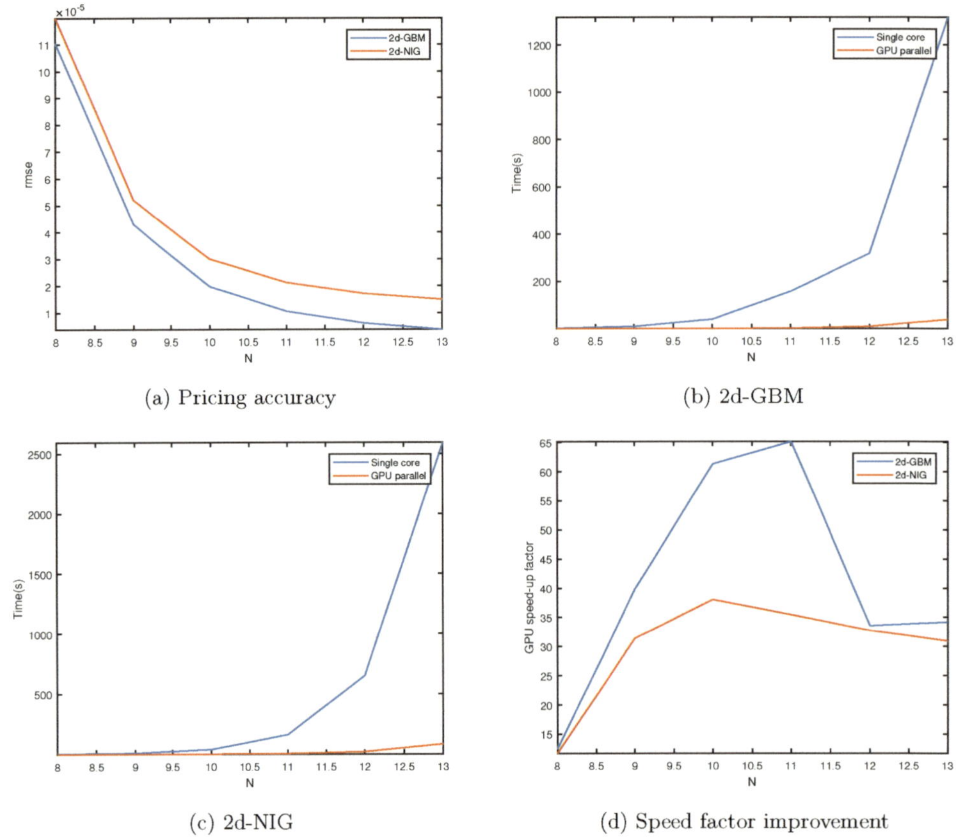

(a) Pricing accuracy

(b) 2d-GBM

(c) 2d-NIG

(d) Speed factor improvement

Figure 1.
Numerical results for spread option pricing.

From **Table 2**, we can see that the optimal values of N (in terms of computation efficiency) are in the middle of the tested range, 9 and 10 for both the 2d-GBM and 2d-NIG models. The decline in performance for larger values of N is due to the increased memory requirements. When compared to 2d-GBM, 2d-NIG seems to have less of an increase in performance when executed on GPU, which could be because it is more computationally heavy (such as calculating the characteristic function $\Phi_{\mathrm{NIG}}(u;T)$, which involves two square root and exponential calculations, as opposed to simply one in the GBM model).

4. Conclusion

In this work, we built on the literature on fast and accurate pricing of spread options based on two-dimensional FFT method using parallel computation. We examined the effectiveness of this approach by comparing the computational times of CPU and GPU implementations of the FFT Spread Option Pricing Algorithm in MATLAB. We have taken benchmarks prices from Monte Carlo simulations with 1000000 paths and 100 discretization time steps. Our results decisively conclude that the execution of the algorithm on a GPU significantly improves computational performance, decreasing the time taken to run by a factor of up to almost 60x. Considering how common spread options are in the financial market, a faster way to price these securities means increased efficiency in transactions involving spread options, and the FFT algorithm implemented for this project also vastly improves the accuracy of spread option pricing. This approach is very useful to accurate calibration of spread options which is recognized to be a challenging exercise.

As an extension to this work, one could develop a 3-asset spread option pricing algorithm using the 3D Fast Fourier Transform Algorithm. Such a scheme, while computationally heavy, could be rendered more efficient by harnessing the power of GPUs through the tools available in MATLAB.

Author details

Shiam Kannan[1] and Mesias Alfeus[2,3]*

1 Ridge High School, NJ, USA

2 University of Wollongong, Wollongong, Australia

3 University of Stellenbosch, South Africa

*Address all correspondence to: malfeus@uow.edu.au; mesias@sun.ac.za

References

[1] Kirkpatrick S, Gelatt C, Vecchi M. Optiminization by simulated annealing. Science. 1983;**1**(23):671-680

[2] Dempster M, Hong S. Spread option valuation and the fast Fourier transform. In: Mathematical Finance-Bachelier Congress. Vol. 2000. 2002. pp. 203-220

[3] Hurd T, Zhou Z. A Fourier transform method for spread option pricing. SIAM Journal on Financial Mathematics. 2010; **1**:142-157

[4] Alfeus M, Schlögl E. On spread option pricing using two-dimensional Fourier transform. International Journal of Theoretical and Applied Finance. 2019;**22**(5)

[5] Hurd T, Zhou Z. A Fourier transform method for spread option pricing. SIAM Journal on Financial Mathematics. 2010; **1**(**1**):142-157

Chapter 9

Use of Transforms in Biomedical Signal Processing and Analysis

Ette Harikrishna and Komalla Ashoka Reddy

Abstract

Biomedical signals like electrocardiogram (ECG), photoplethysmographic (PPG) and blood pressure were very low frequency signals and need to be processed for further diagnosis and clinical monitoring. Transforms like Fourier transform (FT) and Wavelet transform (WT) were extensively used in literature for processing and analysis. In my research work, Fourier and wavelet transforms were utilized to reduce motion artifacts from PPG signals so as to produce correct blood oxygen saturation (SpO2) values. In an important contribution we utilized FT for generation of reference signal for adaptive filter based motion artifact reduction eliminating additional sensor for acquisition of reference signal. Similarly we utilized the transforms for other biomedical signals.

Keywords: Fourier transform, biomedical signals, electrocardiogram signal, photoplethysmographic signal, wavelet transform

1. Introduction

The essence of mathematical design cannot be ignored in the analysis of real world engineering applications i.e. the research in engineering and mathematics is a two way parallel track that interrelates and coordinates towards value added research. In specific, the use of transforms in the field of electrical, electronic and communication engineering is unimaginable. In the present scenario of Covid-19 pandemic, world is looking to sustainable development of biomedical devices for critical monitoring and efficient vaccination for human survival [1–4]. In general, the Fourier transform (FT) is a mathematical tool which transforms the time domain signal into a frequency domain representation used in analysis of biomedical, wireless communication, signal and image processing applications. In literature, many researchers had used this tool in frequency domain analysis of all biomedical signals like electrocardiogram (ECG), photoplethysmographic (PPG) and blood pressure (BP).

In continuation to FT, different transforms were developed to analyze and design of various applications based on the requirement [5, 6]. In general, the FT is used in analysis of stationary signals; the wavelet transform (WT) is a mathematical tool used in analysis of both stationary and non-stationary signals. Discrete wavelet transform (DWT) used in enormous application in various engineering fields.

So, in this chapter we addressed some of the research challenges in ECG and PPG signal processing using Fourier and Wavelet transforms.

IntechOpen

2. Transforms and its use in biomedical signal analysis

2.1 Transforms

Jean-Baptiste Joseph Fourier a French mathematician had developed theoretical and mathematical framework for Fourier analysis and harmonic analysis, which was laid down foundation to other transforms and important applications. Forward and inverse transform of a continuous time Fourier transform (CTFT) for a time domain signal $x(t)$ is defined as

$$X(j\omega) = \int_{-\infty}^{\infty} x(t)e^{-j\varpi t}\,\mathrm{dt} \tag{1}$$

$$x(t) = \frac{1}{2\pi}\int_{-\infty}^{\infty} X(j\omega)e^{j\omega t}\,d\omega \tag{2}$$

Here, $e^{-j\varpi t}$ is called as a basis function for CTFT.

For ease of processing the continuous time signal is converted into discrete time, then the corresponding forward and inverse transform of a discrete Fourier transform (DFT) for a discrete time signal $x(n)$ is defined as

$$X(k) = \sum_{n=0}^{N-1} x(n)e^{-j\frac{2\pi nk}{N}};\ for\ k = 1, 2, 3 \ldots\ldots N \tag{3}$$

$$x(n) = \frac{1}{N}\sum_{k=0}^{N-1} X(k)e^{j\frac{2\pi nk}{N}};\ for\ n = 1, 2, 3 \ldots\ldots N \tag{4}$$

Here, $e^{-j\frac{2\pi nk}{N}}$ is called as a basis function for DFT.

Similarly, the continuous wavelet transform (CWT) is defined as

$$F(a,b) = \frac{1}{\sqrt{a}}\int_{-\infty}^{\infty} \psi\left(\frac{t-a}{b}\right)f(t)\,dt \tag{5}$$

where, $f(t)$ is a time domain signal, $\psi(t)$ is a basis function also called as mother wavelet; F(a, b) is WT of a signal $f(t)$; 'a' and 'b' are shifting and scaling parameters respectively;

Discrete WT (DWT) uses a series of low pass filters (LPF) and high pass filters (HPF) to decompose the signal of interest into different scales as approximate (Aj) and detailed coefficients (Dj). The output coefficients of the LPF are called approximations while the output coefficients of the HPF are called details.

A 3-stage DWT decomposition tree is illustrated in **Figure 1**.

After first level of decomposition A_1 and D_1 will be the outputs, D_1 will be stored in **C** matrix as shown in **Figure 1**. A_1 will be further decomposed in 2nd level of decomposition as D_2, A_2. D_2 will be stored in **C** matrix, and A_2 will be decomposed in 3rd level of decomposition as D_3, A_3. Then A_3 and D_3 will be stored in **C** matrix, i.e. **C** matrix consists of concatenated A_3, D_3, D_2, D_1 coefficients. **L** matrix stores the length of each corresponding approximate and detailed coefficient as in **C** matrix.

C matrix gives the concatenated approximate and detail coefficients. **L** matrix gives length of each coefficient. **C** and **L** matrices along with suitable filters will be used to get back the original time domain signal $f(t)$.

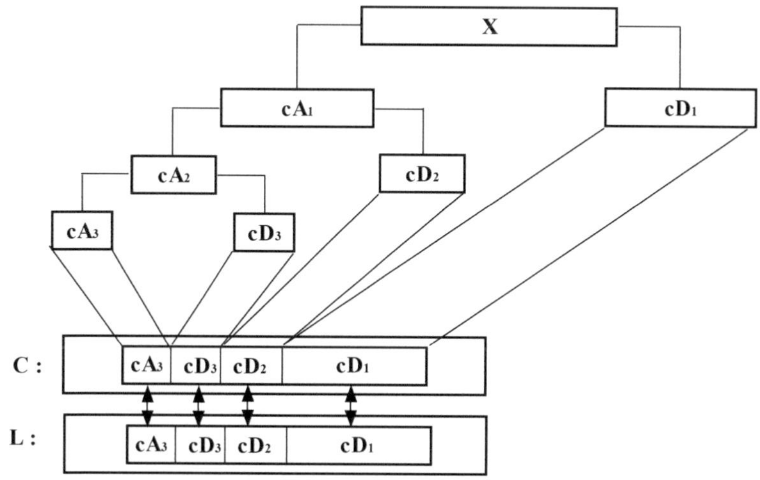

Figure 1.
A sample of 3-stage Wavelet decomposition tree, In Matlab® 'wavedec' command will decompose signal of interest and gives **C** *and* **L** *matrices.*

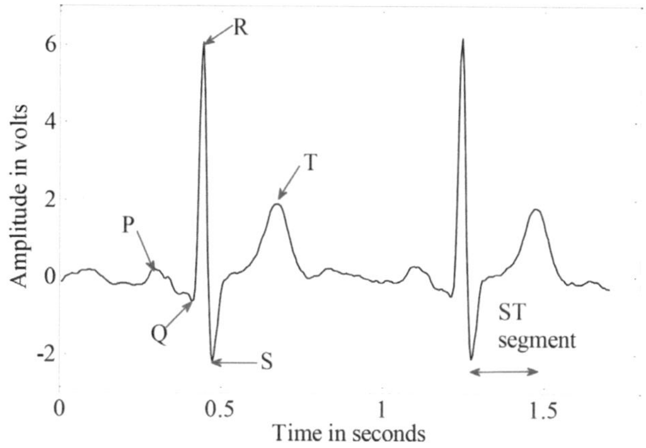

Figure 2.
Typical recorded ECG signal in the laboratory.

2.2 Biomedical signals

The Electrocardiogram (ECG) is an electrical manifestation of contractile activity of the heart, is a representation of instantaneous electrical activity of the heart during its contraction and relaxation over a period of time. A standard 12 leads ECG is recorded with the help of surface electrodes placed on the limbs and chest. In general, ECG is a widely accepted diagnosis tool in clinical validations of heart related diseases. **Figure 2** shows a typical ECG signal, marked with all the characteristic waves and durations such as P wave, QRS complex wave, T wave and ST segment [7].

Monitoring of blood oxygen saturation (SpO_2) is one of the important parameter to know the health status of Covid-19 affected patient. Pulse oximeter uses sensor probes to record photoplethysmographic (PPG) signals so as to estimate the SpO_2 values. Typical recorded PPG signal is shown in **Figure 3**.

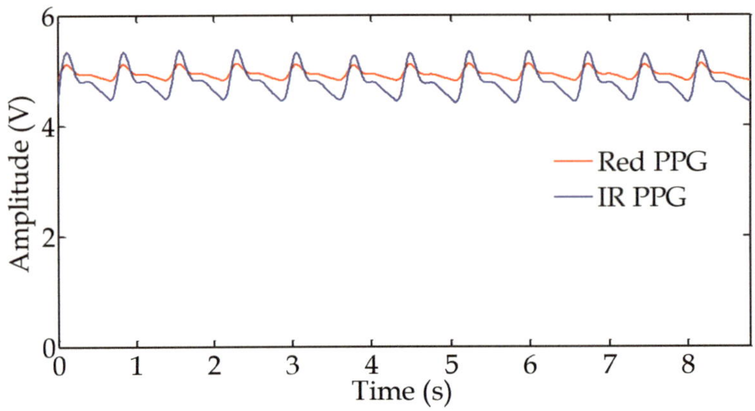

Figure 3.
Typical recorded PPG signal in the laboratory.

2.3 Biomedical signal processing using transforms

Fourier transform (FT) is used to analyze the behavior of biomedical signals in frequency domain. In Matlab FFT command can be used to get the frequency domain signal.

Following is the sample code to plot time and frequency domain signals.

```
load ecg_signal.txt;
ecg_signal_fft=fft(ecg_signal);
figure(1)
subplot(211)
plot(ecg_signal)
subplot(212)
plot(ecg_signal_fft)
```

Some additional modifications will be done in the program to get the following plots.

In the bottom trace of **Figure 4**, it can be seen that the frequency domain components of ECG signal are extending from 0 to 100 Hz, the main source of noise 60 Hz power line interference (PLI) the spike at 60 Hz can be observed. So, the complete frequency domain behavior of signal can be computed using fft command.

Similarly, the frequency domain components of PPG signal is shown in **Figure 5**. It can be seen from the bottom trace that frequency components present in PPG signal are pulse rate or heart rate component and MA noise component. Likewise it can be continued for many biomedical signals to see the frequency domain behavior of the signal.

So, use of Fourier transform in de-noising of signal can be described as shown in **Figure 6**.

In general adaptive filter provide a viable solution when signal and noise are in same frequency range. Adaptive filter requires a two input signals [8].

For example, in a power line interference cancelation from ECG signal, one is recorded noisy ECG signal and other is power line noise. So, here a synthetic noise reference signal will be generated using Fourier transform which will be used as another input to the adaptive filter will potentially eliminate additional sensor for acquisition shown in **Figure 7**.

A synthetic noise reference signal is generated for use in adaptive filtering without using any extra hardware. It is generated from the motion corrupted signal

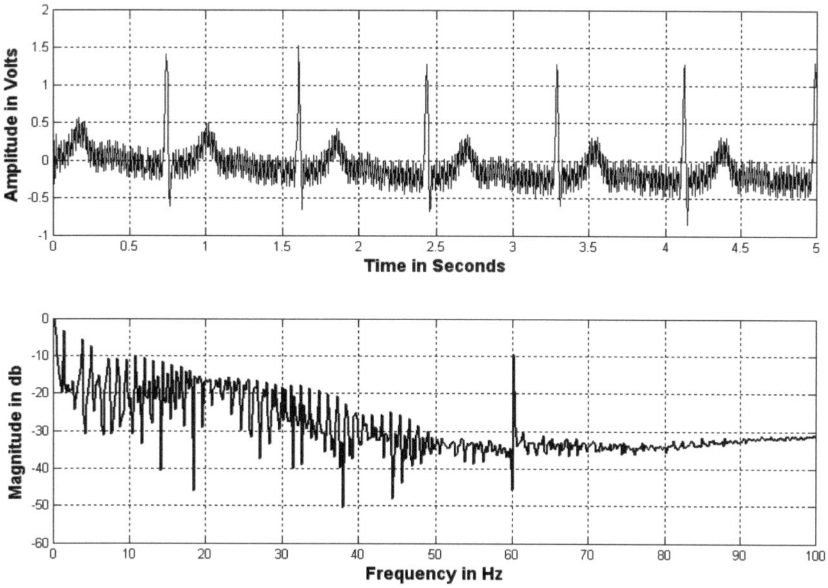

Figure 4.
Recorded ECG signal in top trace and its corresponding spectrum after application of FFT in bottom trace.

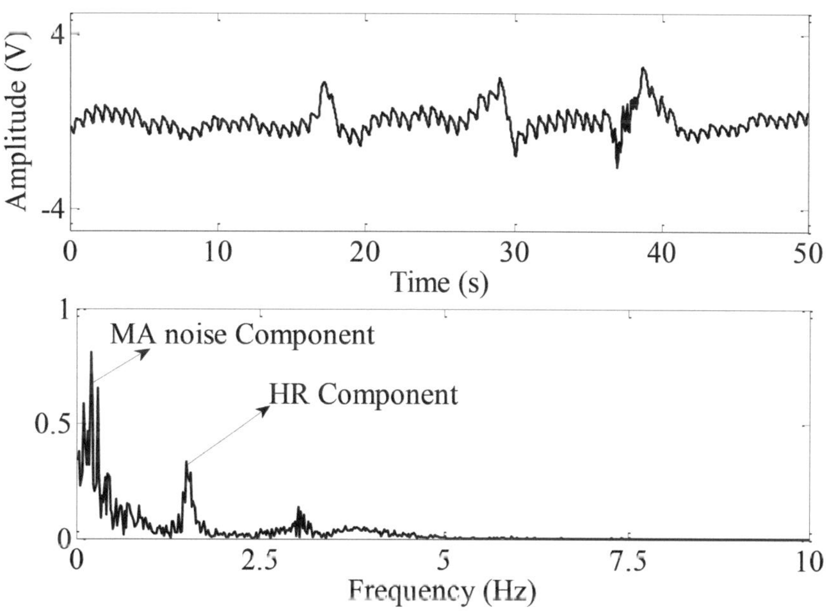

Figure 5.
Recorded PPG signal in top trace and its corresponding spectrum after application of FFT in bottom trace.

in the following way. The frequency spectrum of MA corrupted PPG signal consists of various frequency components, the pulsatile (0.5–4 Hz), respiratory activity (0.2–0.35 Hz) and MA noise component (0.1 Hz or more) information. By setting the co-efficients of cardiac and respiratory activity frequency components in the spectrum of MA corrupted PPG to zero, a modified spectrum corresponding to noise is obtained. By applying inverse Fourier transform to this modified spectrum, a synthetic noise reference signal is generated. The corresponding adaptive filter is shown in **Figure 8**.

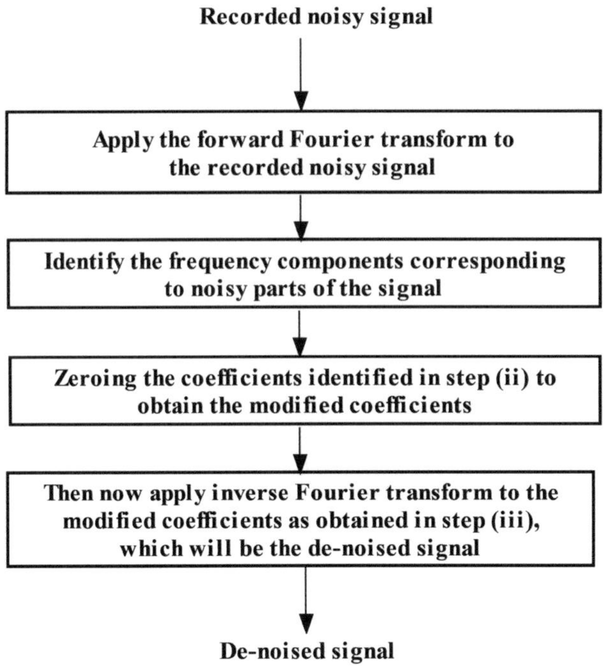

Figure 6.
Flowchart for de-noising of recorded signal using Fourier transform.

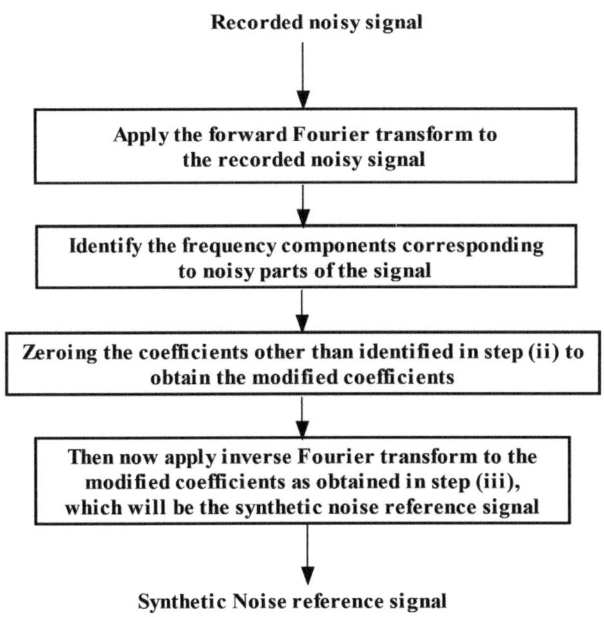

Figure 7.
Flowchart for generation of synthetic noise reference signal using Fourier transform.

With the help of LMS adaptive algorithm, MA noise is removed by estimating the synthetic noise reference signal and adapting the filter coefficients based on filter order. The necessary equations to implement the proposed method are given below:

$$\hat{S}(n) = S(n) + N(n) - \hat{N}(n) \tag{6}$$

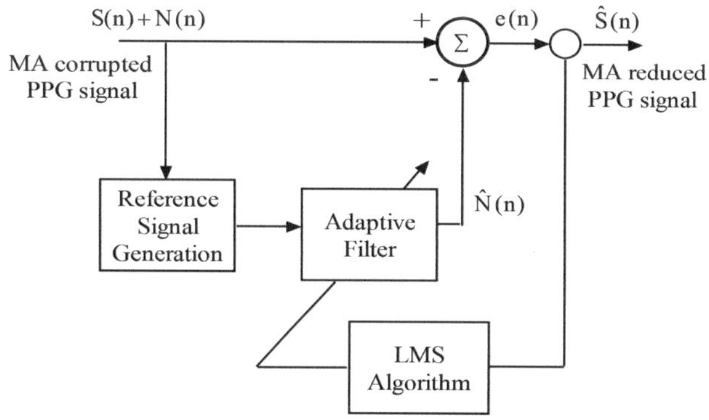

Figure 8.
Motion artifact reduction from PPG signals using Adaptive filter.

$$\hat{N}(n) = \sum_{i=0}^{L} w_i N_R(n-i) \quad (7)$$

$$w_i(n+1) = w_i(n) + 2\mu S(n) N_R(n-i) \quad (8)$$

where i: 0,1,2, … , L, L: filter order, S(n) + N(n): MA corrupted PPG signal, $\hat{S}$(n) =e(n): MA reduced PPG signal, $\hat{N}$(n): estimated synthetic noise reference signal, and NR(n): the synthetic noise reference signal.

The result of above methodology is presented in **Figure 9**, below. **Figure 9(b1)** represents the generated synthetic noise reference signal which potentially eliminates the additional sensor for data acquisition.

The biomedical signals such as ECG and PPG signals are quasi periodic signals i.e. the period of the signal continuously changes with time, but it is a periodic signal. In general, the pure periodic signals are stationary in nature means its period will be constant irrespective of time. So, Fourier transform is not sufficient to analyze the quasi-periodic signals. Wavelet transform will provide a viable solution to the same [9, 10].

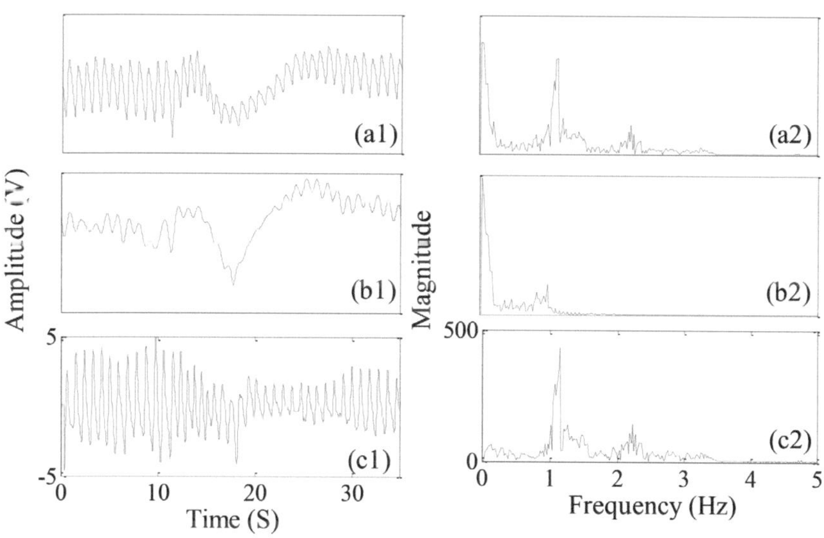

Figure 9.
Recorded PPG signal in (a1), generated MA synthetic reference signal in (b1) and MA reduced PPG signal in (c1) and their corresponding spectra in (a2)-(c2) respectively.

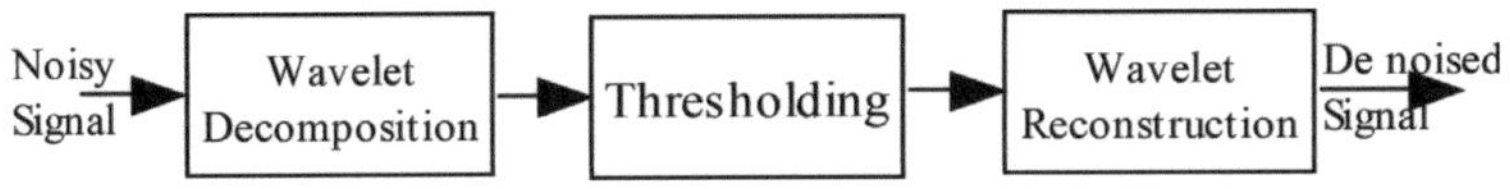

Figure 10.
Wavelet Denoising methodology.

The general de-noising procedure follows the steps described below and shown in **Figure 10**.

Decomposition: *Choose a wavelet and choose a convenient level N for decomposition. Compute the wavelet decomposition of the signal s at level N.*

Thresholding detail coefficients: *For each level from 1 to N, select a threshold and apply soft or hard thresholding to the detail coefficients.*

Reconstruction: *Perform the wavelet reconstruction using the original approximation coefficients of level N and the modified detail coefficients of levels from 1 to N.*

There are two important steps: how to choose the threshold, and how to perform the thresholding [10]. In hard thresholding process, the elements whose absolute values are lower than the threshold will be set to zero. Soft thresholding is an extension of hard thresholding, first setting to zero the elements whose absolute values are lower than the threshold, remaining coefficients are compressed.

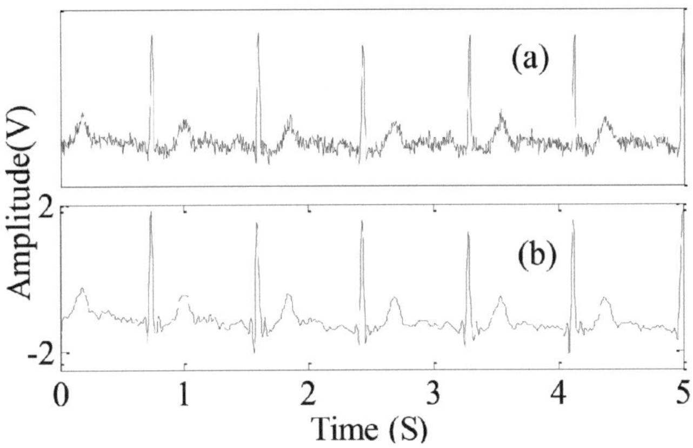

Figure 11.
(a) ECG signal corrupted with Electromyography signal (b) De-noised ECG signal using wavelet de-noising methodology.

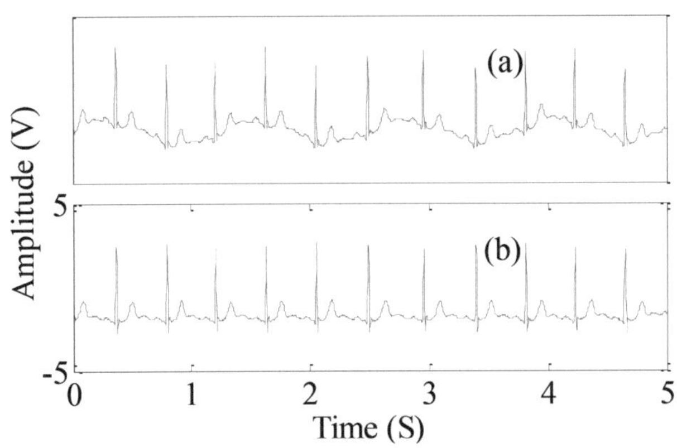

Figure 12.
(a) ECG signal corrupted with baseline noise (b) De-noised ECG signal using wavelet de-noising methodology.

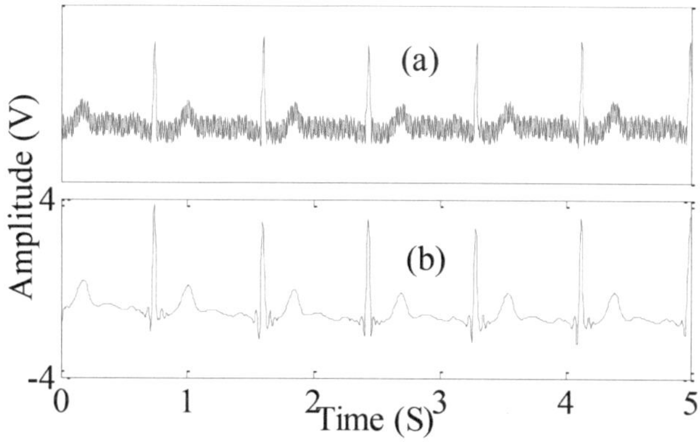

Figure 13.
(a) ECG signal corrupted with power line noise (b) De-noised ECG signal using wavelet de-noising methodology.

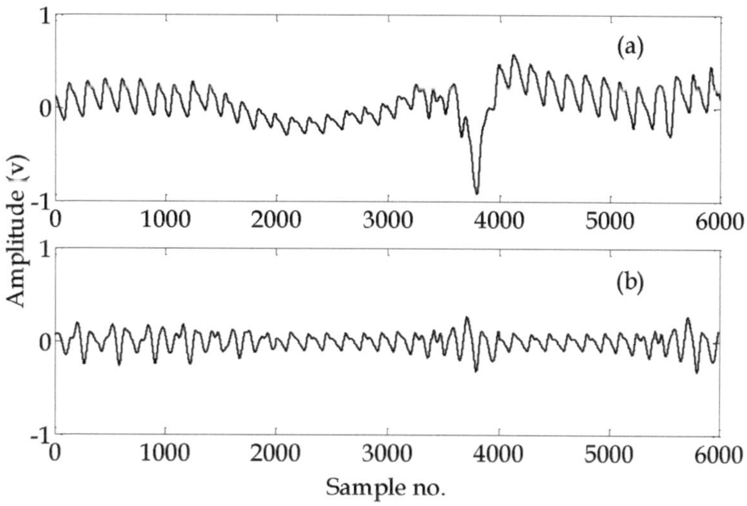

Figure 14.
(a) PPG signal corrupted with MA noise (b) De-noised PPG signal using wavelet de-noising methodology.

The same results were presented in **Figures 11–13**. Similar results were presented for PPG signal as shown in **Figure 14**.

3. Conclusions

Transforms like Fourier and wavelet transforms were used in biomedical signal analysis and processing. Fourier and wavelet transforms were utilized to reduce motion artifacts from PPG signals so as to produce correct blood oxygen saturation (SpO2) values. In an important contribution we utilized FT for generation of reference signal for adaptive filter based motion artifact reduction eliminating additional sensor for acquisition of reference signal.

Author details

Ette Harikrishna[1] and Komalla Ashoka Reddy[2*]

1 Electronics and Communication Engineering, University College of Engineering, Kakatiya University, Kothagudem, Telangana, India

2 Electronics and Communication Engineering, Kakatiya Institute of Technology and Science, Warangal, Telangana, India

*Address all correspondence to: kareddy.iitm@gmail.com

References

[1] Alejandro Domínguez, Alejandro Domínguez: Highlights in the history of the Fourier transform. IEEE Pulse. 2016: 7:1:53-61. DOI: 10.1109/MPUL.2015. 2498500.

[2] Alan V Oppenhiem, Alan S. Willsky, S. Hamid Nawab. Signals and Systems. Prentice Hall, Newjersy, 1983.

[3] I. Daubechies: The wavelet transform, time-frequency localization and signal analysis. IEEE Trans. Inf. Theory, 1990: 130:5: 961-1005. DOI: 10.1109/18.57199

[4] P. M. Bentley and J. T. E. McDonnel: Wavelet transforms: an introduction, Electron. Commun. Eng. J., vol. 6, no. 4, pp. 175-186, Aug. 1994. DOI: 10.1049/ecej:19940401.

[5] Bochner, S.; Chandrasekharan, K. *(1949), Fourier Transforms,* Princeton University Press.

[6] Bracewell, R. N. (2000), The Fourier transform and its Applications (3rd ed.), Boston: McGraw-Hill, ISBN 978-0-07-116043-8.

[7] P. S. Hamilton, “A comparison of adaptive and non-adaptive filters for the reduction of PLI in the ECG,” IEEE Trans. Biomed. Eng., vol. 43(1), pp. 105-109, 1996.

[8] Lino Garcia Morales, Adaptive filtering, Intech open, DOI: 10.5772/675.

[9] J.S Sahambi, S.N. Tandon and R.K.P. Bhatt, “Using Wavelet Transform for ECG Characterization,” IEEE Eng. in Med. and Bio., 1997.

[10] D. L. Donoho, De-noising by softthresholding, IEEE Transaction on Information Theory, Vol. 41, pp. 613–627, May 1995.

Chapter 10

Insights from Systematic DFT Calculations on Superconductors

Ian D.R. Mackinnon, Alanoud Almutairi and Jose A. Alarco

Abstract

We present three systematic approaches to use of Density Functional Theory (DFT) for interpretation and prediction of superconductivity in new or existing materials. These approaches do not require estimates of free parameters but utilize standard input values that significantly influence computational resolution of reciprocal space Fermi surfaces and that reduce the meV-scale energy variability of calculated values. Systematic calculations on conventional superconductors show that to attain a level of resolution comparable to the energy gap, two key parameters, Δk and the cut-off energy, must be optimized for a specific compound. The optimal level of resolution is achieved with k-grids smaller than the minimum reciprocal space separation between key parallel Fermi surfaces. These approaches enable estimates of superconducting properties including the transition temperature (T_c) *via* (i) measurement of the equivalent thermal energy of a phonon anomaly (if present), (ii) the distribution of electrons and effect on Fermi energy (E_F) when subjected to a deformation potential and (iii) use of parabolic, or higher order quartic, approximations for key electronic bands implicated in electron–phonon interactions. We demonstrate these approaches for the conventional superconductors MgB_2, metal substituted MgB_2 and boron-doped diamond.

Keywords: Fermi energy, Fermi level, Fermi surface, reciprocal space, density functional theory, parabolic equations, phonon dispersions, transition temperature, magnesium diboride

1. Introduction

Design and synthesis of new materials requires translation of reciprocal space detail in electronic band structures (EBSs) and phonon dispersions (PDs) to equivalent real space representations [1–4]. We have shown that for conventional superconductors (SCs), the format and depth of modes in PDs associated with a Kohn anomaly are strongly influenced by the computational resolution of DFT models [5]. Our view is that EBS calculations, performed with appropriate resolution, may also provide critical information on the superconducting gap, which for many SC materials, is in the meV range.

The Fermi level, Fermi energy (E_F), and the density of nearly free-electron carriers calculated by DFT are key values that, to date, have been variously reported with differences of many hundreds of meV for the same compound. For example, the value of E_F for a well studied compound such as MgB_2 has been variously reported as "several eV" [6], 0.55 eV or 0.122 eV [7], and more recently, as 0.428 eV [8]. In comparison, many publications on MgB_2 and software packages such as

CASTEP and ADF consistently report a value for E_F of ~8.4 eV [9–11]. Calculated variations of this magnitude have garnered limited attention [8] for SCs due to an underlying assumption that $E_F/k\theta >> 1$ (where k is Boltzman's constant and θ is the Debye temperature). Due to this assumption and the deceptive influence of average values for phonon frequencies, a value for E_F is not considered in the simplified McMillan version of the Eliashberg model for superconductivity [12]. However, as noted by Malik [8], equations that explicitly include E_F and/or critical current (j_o) values may provide clues on how to increase or modify T_c. Malik suggests that regardless of physical attributes, SCs may be distinguished by their values of E_F [8].

Approximations to the Eliashberg model that minimize computational cost require estimates of the electron–phonon interaction, λ, and the Coulomb strength, μ^* [13, 14]. For many conventional SCs, these parameters are limited to a narrow range of values and provide reasonable estimates for superconducting properties of known materials [15–18]. More recently, Sanna *et al.* [19] show that a fully *ab initio* Eliashberg approach provides good estimates of superconducting properties including T_c for a range of compounds without invoking estimates of free parameters such as μ^*. This work by Sanna *et al.* [19], as well as development and use of the Superconducting Density Functional Theory (SCDFT) [20, 21], are elegant computational approaches to the Eliashberg model that have successfully predicted superconducting properties of new materials such as H_3S [19, 22]. Nevertheless, these codes are not universally available to materials researchers particularly if deep mathematical rigor is required for implementation.

In our search for new SCs, we have evaluated the computational resolution and electron–phonon detail possible with DFT codes readily available in well-known software packages such as CASTEP or Quantum Espresso, to name a couple of examples. Using MgB_2 and similar Bardeen-Cooper-Schrieffer (BCS) compounds, we have systematically explored the sensitivity and use of PD and EBS constructs to calculate key superconducting properties without recourse to free parameter estimates or modification of functionals. We initially explored use of a phonon anomaly to estimate T_c [23, 24]; an approach that appears effective for strong phonon mediated superconductivity including for metal substituted MgB_2 [25]. In other work, we extended this systematic approach to evaluate PDs and EBSs for a wide range of metal diboride compounds (*e.g.* ScB_2, YB_2, TiB_2) using DFT at appropriate computational resolution [26].

More recently, we have examined the link between PDs and EBSs and, in particular, the topology of the Fermi Surface (FS) with pressure [27] and the change in electron density distributions as MgB_2 transitions to the superconducting state [28]. In both approaches, we are able to confirm experimentally determined superconducting properties for a range of conventional (BCS) compounds and then, to predict T_c for new metal substituted analogues of MgB_2 [23, 25]. These approaches are, in essence, empirical methods, which systematically identify regular dispersion patterns in calculated PDs and EBSs, based entirely on accepted codification of the DFT [9, 10] and, equally, a clear understanding of input parameter limitations that determine computational resolution. Thus, this use of DFT software, underpinned by elegant formalism and constructs by Kohn and colleagues [29, 30], is a complement to approximations of the Eliashberg model [19, 20].

In this work, we show why computational resolution for DFT models of EBSs and PDs for conventional SCs is critical. In addition, we delineate a third approach to estimate the superconducting gap using parabolic, or higher order quartic, approximations to key bands in the EBS. This approach requires examination of an extended Brillouin zone (BZ) schema and demands lower computational cost compared to equivalent PD calculations for similar outcome(s). When applied to MgB_2, at sufficiently fine k-grid and reciprocal space cut-off, this approach directly estimates the superconducting gap and assists identification of valence bands and the origin for E_F. In combination, these three approaches provide reliable property predictions for unknown, or theoretical, structures for materials researchers.

2. Calculation methods

EBS calculations are undertaken using DFT as implemented in the Cambridge Serial Total Energy Package (CASTEP) of Materials Studio (MS) 2017 and 2018 [9, 10]. All structures are optimized for geometry, including cell parameters, starting with crystal information files (.cif) available in standard databases. In general, the local density approximation (LDA) and generalized gradient approximation (GGA), with norm-conserving pseudopotentials, are used in DFT calculations. The typical setup for calculations uses a k-grid ranging from 0.06 $Å^{-1}$ to 0.02 $Å^{-1}$ or smaller, with a plane wave basis set cut-off of 990 eV, ultra-fine (or better) customized setup to ensure total energy convergence of less than 5×10^{-6} eV/atom, a maximum force of less than 0.01 eV/Å, a maximum stress of less than 0.02 GPa and maximum displacements of less than 5×10^{-4} Å.

All outputs meet convergence criteria at the same fine tolerance level for geometry optimization. The effects of input parameters to DFT calculations described in this work are not related to differences in calculation convergence, but critically, are due to the discreteness, or the finite number of the reciprocal space points, used to select plane waves as basis functions. To illustrate particular points, we also vary specific parameters such as basis set cut-off values, Δk values or software versions as identified in the text.

We also perform numerical interpolations of EBSs to validate higher order trends and to delineate fine structure in computed outcomes. To obtain parabolic, or higher order polynomial, approximations to the electronic bands, the DFT calculated data from MS is exported in csv/excel format. For MgB_2, sections of particular bands in the energy range − 14 to 4 eV along the Γ-M (and Γ-K) directions are selected and mirrored across the vertical axis at Γ. Individual parabolic, or higher order quartic, trendline fittings are obtained and used to overlay for comparison with the (periodically repeated) extended BZ scheme of the DFT calculated EBS. Effective masses are also calculated and evaluated for parabolic approximations of different branches of the EBS.

3. Computational resolution

We provide outputs from a series of *ab initio* DFT calculations on two SC compounds with substantially different experimentally determined T_c values (*i.e.* MgB_2 $T_c \sim 39.5$ K; B-doped diamond $T_c \sim 4.0$–7.5 K depending on level of doping) [31–33]. For both compounds, when the value of k-grid is varied in the examples below, all other parameters are maintained the same for all calculations. A range of k-grid values are exemplified in order to highlight differences in sensitivity of EBS and PD outputs for SC compounds. These examples highlight the key role computational resolution can play with interpretation of SC properties.

3.1 Band structure – variation with k-grid

We have been intrigued by the potential to directly determine the superconducting gap energy for a BCS SC using an appropriate resolution EBS. In this regard, MgB_2 offers good opportunity to evaluate this potential due to well defined crystallography and the key role ascribed to σ bands and superconductivity [12, 34]. DFT calculations for a compositional suite, or structure type, produce EBSs that show, in general, similar formats even when two different functional approximations are used [26]. A typical outcome for MgB_2 using the LDA approximation in the CASTEP module of Materials Studio using a k-grid value of 0.008 $Å^{-1}$ is shown in **Figure 1**. These band structures convey useful information for elucidation of potential for superconductivity. For example, the σ bands appear as approximate inverted parabolas (in red and blue lines; green dotted box) near the Γ center point and cross the Fermi level on either side of Γ (**Figure 1**). These bands display a strong electron–phonon coupling to the E_{2g} phonon modes and are implicated in superconductivity for MgB_2 *via* both theoretical and experimental analyses [35–38].

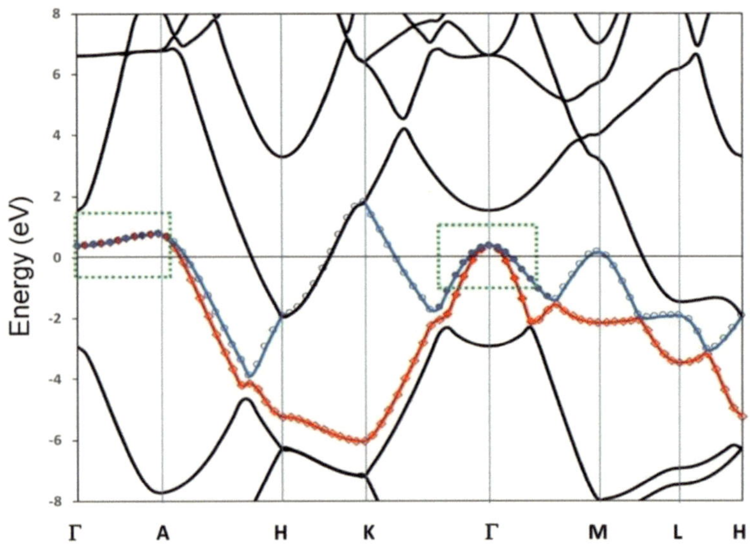

Figure 1.
Electronic band structure for MgB_2 calculated with the LDA approximation for k = 0.008 $Å^{-1}$ using the CASTEP module of materials studio [9]. The green boxes enclose sections of σ bands.

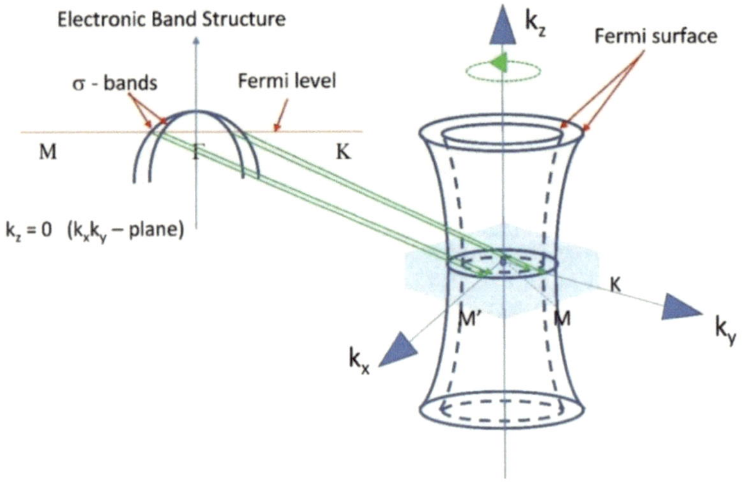

Figure 2.
Schematic showing the direct relationship between σ bands crossing the Fermi level around Γ and the topology of the FS.

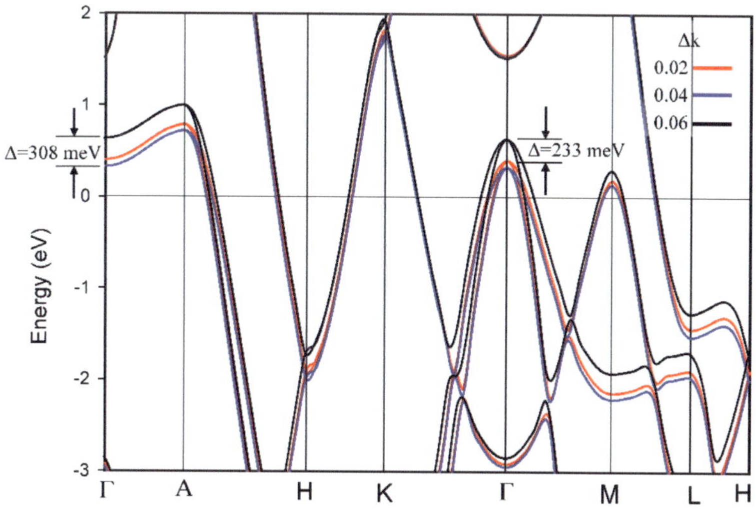

Figure 3.
Electronic band structure for MgB_2 calculated with the LDA approximation for k-grid values 0.02 $Å^{-1}$ (red), 0.04 $Å^{-1}$ (blue) and 0.06 $Å^{-1}$ (black) using the CASTEP module of materials studio. Notice the substantial differences in band energies, particularly in regions associated with the σ bands.

For MgB_2, the relationship of these σ bands near the Fermi level on either side of Γ is shown in the schematic in **Figure 2**. The reciprocal space projection of degenerate σ bands at the Fermi level correspond to the three-dimensional reciprocal space representations of two FSs at Γ parallel with the k_z direction. That is, except for the direction along k_z, the band is a 2D projection at $k_z = 0$ of the 3D Fermi surface representation. The region between these two FSs is sensitive to the reciprocal space projection(s) at the Fermi level and, as noted in previous work, is key to the superconducting mechanism in MgB_2 [27, 28].

Variations in the calculated energies of specific bands, for example in MgB_2 where electron–phonon coupling is predominantly linked to the σ bands [17, 39, 40], are strongly influenced by the k-grid value used in DFT computations. **Figure 3** demonstrates the effect of k-grid value, or the sensitivity of DFT calculations, on the EBS for MgB_2 using the LDA functional for a series of Δk values 0.02 $Å^{-1}$, 0.04 $Å^{-1}$ and 0.06 $Å^{-1}$. The k-grid value affects calculated energies for bands near the Fermi level particularly those σ bands associated with superconductivity highlighted in **Figure 1** for MgB_2. The differences in energy at Γ or A between calculations range from tens of meV to hundreds of meV for three different k-grid values as shown in **Figure 3**.

In **Table 1**, we show substantial meV shifts in enthalpy and E_F for MgB_2 calculated at different k-grid values using the same functional and the same ultra-fine tolerance for geometry optimization convergence. For MgB_2, **Table 1** shows the difference in energy, ΔE_v (in eV), between the Fermi level and the vertex of the parabola at Γ for different values of Δk. Differences in lattice parameters (*i.e.* ~0.01 Å), enthalpy values (*i.e.* ~10–20 meV) and E_F values (*i.e.* ~200 meV) are evident for geometry optimized calculations with different k-grids. **Table 1** also shows that as the Δk value is reduced, values for enthalpy achieve a consistent value for MgB_2.

We also show calculated values for B-doped diamond using different k-grid values in **Table 1**. In this case, k-grid intervals are smaller (0.005 $Å^{-1}$ < Δk < 0.020 $Å^{-1}$) than those used for MgB_2 with corresponding smaller shifts in enthalpy and E_F. This lower magnitude impact of the k-grid is in part due to fewer degrees of freedom (*e.g.* cubic symmetry compared to hexagonal) and a significantly lower value for T_c [32, 33], with corresponding FSs in closer reciprocal space proximity. Nevertheless, the E_F value

Compound	k-grid value [Å^{-1}]	Lattice parameters [Å] *a*	*c*	Enthalpy [eV]	Fermi energy [eV]	ΔE_v [eV]
MgB_2 $T_c \sim 40$ K	0.010	3.038657	3.487973	−1748.5037	8.4055	0.3424
	0.018	3.038630	3.487921	−1748.5039	8.4126	0.3975
	0.020	3.039107	3.486623	−1748.5046	8.3976	0.4038
	0.040	3.042948	3.478367	−1748.5003	8.4704	0.3281
	0.060	3.031477	3.514558	−1748.4685	8.2108	0.6365
B-doped diamond $T_c \sim 4$–7.5 K	0.005	3.582982		−1163.6112	11.0832	1.6164
	0.010	3.582982		−1163.6114	11.0844	1.6152
	0.015	3.582984		−1163.6121	11.0956	1.6039
	0.017	3.582987		−1163.6120	11.0941	1.6056
	0.020	3.583024		−1163.6102	11.0579	1.6410

Table 1.
Parameters calculated for MgB_2 and for B-doped diamond.

differs by ~35 meV and the difference in energy to the vertex of the band at Γ (*i.e.* ΔE_v) differs by up to ~35 meV depending on the Δk value.

We show a more detailed systematic comparison of calculated enthalpies as a function of the k-grid value for MgB_2 in **Figure 4**. As noted above, all calculations are converged to the same ultra-fine criteria, or tolerance, where self-consistency is achieved. The variability of results shown in **Figure 4** is due to the discreteness of functions and values used to derive the full solution of the Schroedinger equation. The variability is not due to lack of convergence which in all cases is defined in Section 2 above.

Figure 4a shows that the value of enthalpy for MgB_2 oscillates around a consistent minimal value of −1,748.502 eV as the k-grid value is decreased to <0.015 Å^{-1}. The context for this variation in enthalpy is shown in **Figure 4b** where the k-grid value is extended to 0.2 Å^{-1}– a value that has been used in some DFT calculations as criterion for machine learning algorithms [41]. At these higher values for Δk, enthalpy calculations do not provide useful information on subtle structural

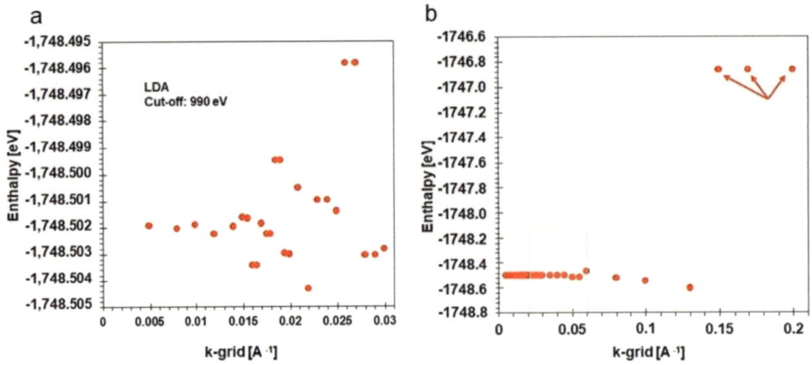

Figure 4.
*Systematic comparison of calculated enthalpies for MgB_2 (a) for fine values of k-grid (i.e. < 0.03 Å^{-1}) and (b) for coarser grid values including those utilized for machine learning searches (arrowed) of materials databases [41]. Enthalpies shown in Figure 4(a) are reproduced in (b) for reference. The lightly shaded region in **Figure 2(b)** delineates the k-grid values used for EBS calculation in **Figure 3**.*

variations due to superlattices or of order/disorder. For example, we have shown using DFT calculations with appropriate k-grid values for $(Mg_{1-x}Al_x)B_2$ that ordered motifs with adjacent Al-layers are thermodynamically favored by ~0.15 eV over more complex, disordered configuration(s) [42]. A discrepancy of ~0.2 eV due to incorrect choice of k-grid value (**Figure 4**) does not enable such distinction to be made with confidence.

3.2 Influence of atom displacements

In a dynamic system, other factors may also influence the position of key electronic bands with respect to the Fermi level. For example, atoms in all solids at temperatures above absolute zero vibrate [43] and in some cases, the resulting phonons may align with specific crystallographic real space features such as inter-atom bonds. This circumstance occurs for MgB_2 in which one of the dominant E_{2g} phonon modes – shown to be intimately involved in electron–phonon coupling at the onset of superconductivity [17, 34] – aligns with B–B bonds in the *ab* plane [36]. Using DFT, we can model the effect of bond deformation along specific planar orientations by displacing atoms from their structural equilibrium positions consistent with the direction of the E_{2g} phonons [26, 28]. Under different extents of displacement, electron density distributions along the B–B bond and the corresponding EBS, can be determined [28, 36].

Figure 5 shows the effect on the EBS for MgB_2 of atom displacement along the B–B bond by ~0.6% (*i.e.* a shift of ~0.063 Å) from equilibrium [28]. **Figure 5** shows that the E_{2g} phonon, which is degenerate at Γ with a peak parabola at 398 meV, splits into two separate non-degenerate bands above and below the equilibrium condition. The upper σ band - which we attribute to the heavy effective mass - has a calculated energy 813 meV above the Fermi level. Thus, parallel or nearly parallel FSs attributable to the superconducting condition [44], no longer exist with a 0.6% shift in atom position(s) [28]. A shift of atom position(s) is also reflected in the form and energy of key phonon modes in the corresponding PD for MgB_2 [26]. An atom displacement of 0.6% along B–B for MgB_2 is not unreasonable at temperatures >40 K [28].

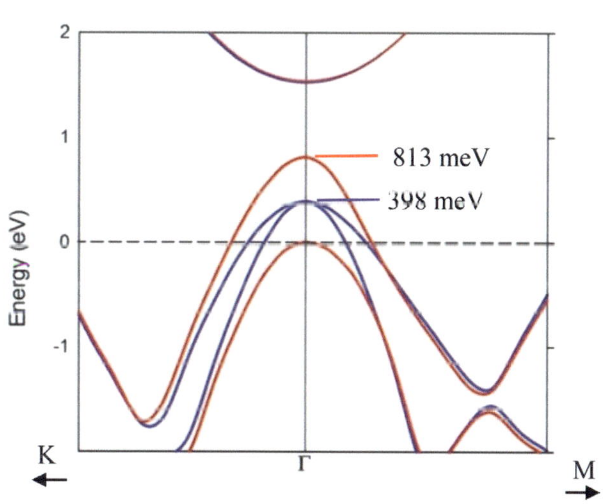

Figure 5.
Enlarged view of EBS around Γ for MgB_2 using LDA functional and Δk = 0.018 $Å^{-1}$ showing (a) degenerate σ bands at equilibrium (blue lines) and (b) after atom displacement D_x = 0.063 Å (red lines) along the E_{2g} mode direction; note the split of σ bands causing loss of degeneracy which coincides with loss of superconductivity [26].

The σ bands for MgB_2 consist of two bands degenerate at the Γ point, but degeneracy is lost when the vector k does not equal 0. The two bands thus have different effective masses (or curvatures) which appear to vary as a deformation potential is applied. In fact, the curvatures become very similar and/or identical along Γ–K at the point where a deformation shifts one band tangential to the Fermi level as shown in **Figure 5**. Along Γ–M, the effective masses are shown to cross over, where the top band has the curvature of the original light effective mass band, and for k-vectors away from the origin, the curvature remains the same as the original heavy effective mass band.

The response of the effective masses (*i.e.* the σ bands) to the deformation potential suggests that electronic behavior associated with superconductivity may be explained by directly analyzing the reciprocal space trajectories of these effective masses. Such an interpretation depends on the collective electronic response in reciprocal space which may, by inference, transform into localized information in real space. This outcome is, again, strongly dependent on the use of a fine k-grid value (< 0.015 Å^{-1}) wherein a polynomial approximation to the bands can be calculated [45]. **Figure 6** shows this polynomial calculated as an average band (dotted line) of the two σ bands for the EBS of MgB_2 at an equilibrium position and for the degenerate condition arising from a deformation potential shift of 0.063 Å along the B–B bond.

The calculated coefficients for the polynomials, fitted to σ_H, σ_L and the average trend line for these bands, σ_M, are dependent on the k-grid values used in DFT calculations as shown for the equilibrium condition (**Figure 6a**) in **Table 2**. The terms of these polynomial coefficients (*i.e.* for X^0 in **Table 2**) for the σ_H and σ_L bands show an interesting characteristic with k-grid value. For example, if we assume that the term is in eV and describes the intersection of the polynomial trend curve with the vertical axis (*i.e.* along the y axis), then the difference between coefficients for the σ_H and σ_L bands not only varies with k-grid value but also approximates the SC gap energy for MgB_2 at finer k-grid values [12, 34, 46]. As the k-grid value increases, the notional SC gap energy also changes and is zero for $\Delta k = 0.03$ Å^{-1}. This trend supports the notion that a higher resolution DFT calculation (*i.e.* with $\Delta k < 0.02$ Å^{-1}) may provide indicative energy gap values for an SC compound directly from an EBS calculation [45]. **Table 2** also shows coefficients for the polynomials at the displaced condition with boron atom positions along the E_{2g} mode direction [28] calculated for different values of Δk. As expected, with σ_H

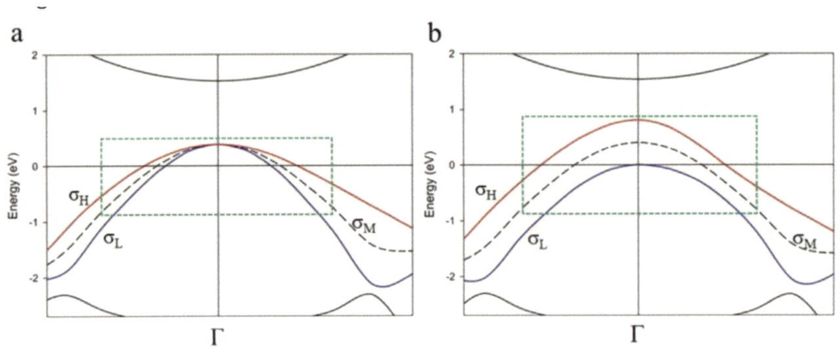

Figure 6.
Sigma bands (heavy: Red, light: Blue and average: Black dotted) for the EBS of MgB_2 along the Γ–K and Γ–M directions calculated with the LDA functional using $\Delta k = 0.01$ Å^{-1} (a) for equilibrium boron atom positions and (b) for degenerate bands formed by displaced boron atom positions along E_{2g} mode directions by 0.6% relative to the equilibrium position. The green dotted region delineates extent of polynomial trend lines matched to the EBS.

Grid Value ($Å^{-1}$)	Coefficient X^4		Coefficient X^2		Coefficient X^0		Coefficient X^0
	σ_H^*	σ_L^*	σ_H^*	σ_L^*	σ_H^*	σ_L^*	σ_H- σ_L (eV)
At Equilibrium (D_x = 0.0)							
0.005	4.2955	18.572	8.161	14.749	0.3441	0.3266	0.0175
0.008	8.0342	36.301	11.159	20.344	0.3441	0.3287	0.0154
0.010	4.0135	17.807	7.960	14.371	0.3912	0.3745	0.0167
0.022	6.8879	36.014	10.381	19.477	0.3377	0.3289	0.0088
0.030	6.4614	42.492	10.330	20.213	0.3830	0.3830	0.0
Displaced along E_{2g} (D_x = 0.006)							
0.005	382.94	36.37	46.15	25.80	0.753	−0.032	0.785
0.008	382.94	36.37	46.15	25.80	0.753	−0.032	0.785
0.010	321.72	19.84	44.98	26.12	0.751	−0.031	0.784

$^*\sigma_H$, σ_L are the heavy and light sigma bands, respectively.

Table 2.
Calculated polynomial coefficients for sigma bands along Γ–K for MgB_2.

and σ_L tracing parallel bands on either side of Γ, along Γ–K the linear terms show a large energy gap, almost constant as a function of Δk, of ~0.78 eV (**Figure 6b**).

3.3 Brillouin zone schemes and high order quartic approximations

Conceptually, it is generally accepted that two approaches: (a) the free-electron theory and (b) the tight binding, or linear combination of atomic orbitals (LCAO), offer reasonably good approximations to the conduction and valence bands, respectively, in the electronic structure of materials [1, 2, 47–50]. With increased computational power, the distinction between these two models becomes negligible. In general, the actual EBS should be similar to an average of respective contributions from these two types of approximations.

The detailed origin of particular bands and that of the zero of E_F can be more readily appreciated when we examine an extended Brillouin zone scheme, instead of a reduced zone scheme [1, 2, 48, 51, 52]. For example, **Figures 7a-c** show periodically repeated reciprocal unit cells with reference to the extended Brillouin zone schemes for the electronic bands of MgB_2 along the Γ-M and Γ-K directions, respectively. **Figure 7a** shows a 2D representation of multiple reciprocal unit cells viewed along c* and identifies reciprocal directions for the calculated EBSs for extended BZs in **Figure 7b** and **c**.

The origin of E_F determined by the parabolic approximation is identified at (i) the cross-over of two parabolas "1" at M in **Figure 7b** and (ii) the inflection point of parabola 3 at K + M in **Figure 7c**. The calculated E_F for MgB_2 at zero pressure using the LDA functional in CASTEP (and Δk ~0.01A^{-1}) is 8.4055 eV. **Figure 7b** and **c** show the location for the origin of E_F, at a K + M type reciprocal space position or the midpoint between two reciprocal space Γ vectors along Γ–K. This location is difficult to infer from a reduced BZ scheme, particularly for complex structures. Nevertheless, these two locations at M and ± K ± M nodal points, directly relate to the real space B–B hexagonal plane in the MgB_2 structure.

For MgB_2, sections of the σ bands along the Γ–M and Γ–K directions are approximated by upward facing parabolas, even when inside the valence band region.

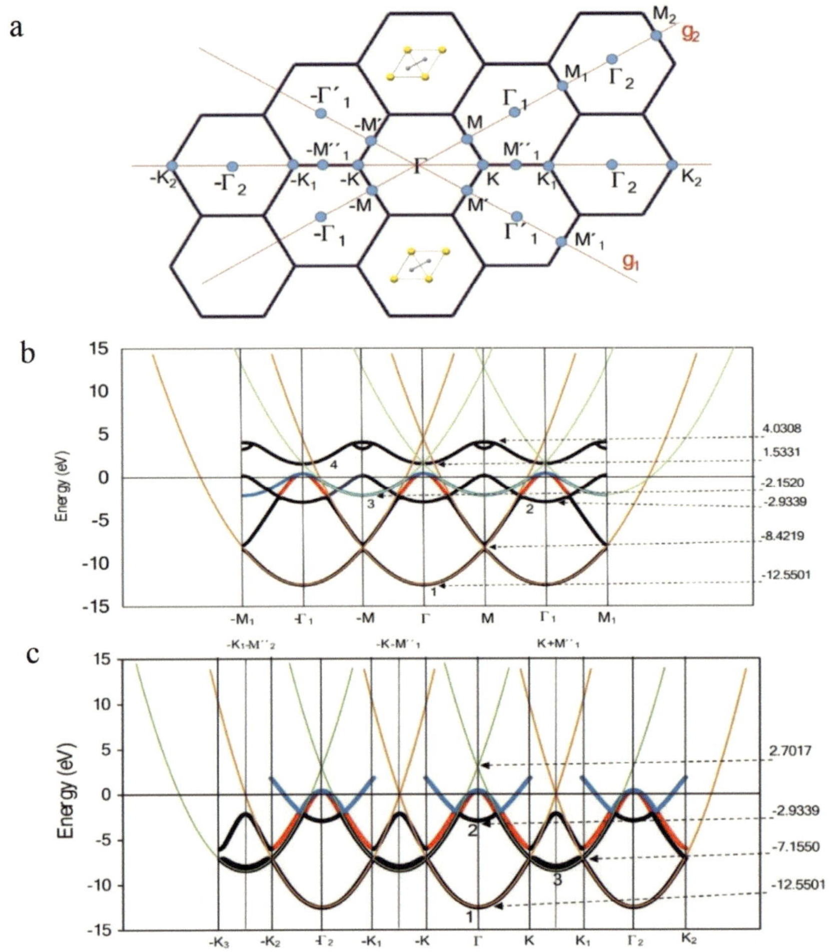

Figure 7.
(a) Schematic of Brillouin zones for MgB_2 viewed along c with nodal point nomenclature for primary reciprocal space orientations along Γ-M and Γ-K. This schematic shows the orientation of the real space asymmetric unit (Mg atoms are yellow; B atoms are gray) as well as nodal points and zones identified in Figure 7b and c. Extended Brillouin zone schemes for the EBS of MgB_2 along: (b) Γ-M and (c) Γ-K. Representative values of energy at the zone boundary and at zone centres are indicated. Energy band sections are labeled as types 1 to 4. The red and blue lines in both EBS schemes refer to similar traces in **Figure 1**.*

Deviations from parabolas occur particularly at zone boundaries where the periodic crystal potential primarily influences free-electron like level crossings [49, 52, 53]. Along Γ–M, a parabola with convex inflection at Γ occurs at −12.55 eV (parabola 1, **Figure 7b**) and its translated homologs reproduce large sections of the light effective mass σ band distant from Γ.

Similarly, a parabola with convex inflection at M and at -M at −2.152 eV (parabola 3, **Figure 7b**) reproduces the heavy effective mass σ band. Along Γ–K, differentiation of the σ bands is less pronounced in the extended zones but is apparent at Γ (**Figure 7c**). In **Figure 7c** along Γ–K, an additional K–M section is shown because a hexagonal boundary edge, equivalent to K–M by symmetry, transects an adjacent reciprocal space point outside the first BZ (node M_1 in **Figure 7a**). Both **Figure 7b** and **c** show the origin of E_F for this structure. **Table 3** summarizes values of key parameters associated with these parabolic approximations to extended BZ schemes for the Γ–M and Γ–K directions, respectively.

Orientation	Band Type	Energy at Γ (eV)	Energy at M (eV)	Energy at K (eV)	Effective Mass
Γ–M	1	−12.5501	−8.4219		1.2821
	2	−2.9339	0.17329		1.3145
	3	0.3979	−2.1520		1.4598
	4	1.5331	4.0308		1.3236
Γ–K	1	−12.5501	—	−7.1550	1.3059
	2	−2.9339	—	—	1.2900
	3	—	−8.4200	−7.1550	1.4311

Table 3.
Calculated parameters at Γ and M, K points for MgB_2.

3.4 Fermi energy values

The value of E_F for a particular DFT calculation is not only sensitive to the k-grid value as shown above but also to other extrinsic conditions such as compositional substitution in a solid-solution and/or changes in applied pressure. **Figure 8** displays the calculated Fermi energies for Al and Sc substitution in $Mg_{1-x}Al_xB_2$ and $Mg_{1-x}Sc_xB_2$ determined with the LDA and GGA functionals. For each calculated series, the value of k-grid is constant (i.e. $\Delta k = 0.02$ Å^{-1}). **Figure 8** also shows the calculated Fermi energies for other end-member compositions with AlB_2-type structure. Note that NbB_2 and ZrB_2 are reported superconductors at very low temperatures ($T_c = 0.6$ K and 5.5 K, respectively) albeit non-stoichiometric or substituted niobium diboride (*e.g.* $NbB_{2.5}$ or $Nb_{0.95}Y_{0.05}B_{2.5}$) is superconducting at 6.3 K and 9 K, respectively [54].

Table 4 lists key parameters based on EBS calculations for MgB_2 with applied external pressure. For each calculation, the LDA functional and k-grid value is constant and values are computed after geometry optimization. The value for the effective mass, m_{eff}^{H}, is determined from the parabolic approximations described below in Section 3.4.

Table 4 shows that as pressure is applied, E_F largely conforms to the textbook equation:

$$E_F = \frac{\hbar^2}{2m_{eff}}\left(\frac{n}{V_0}\right)^{2/3} \tag{1}$$

by using the volume and heavy effective mass obtained at a particular external pressure (assuming that the electron density n does not change with pressure [55] as shown in Eq. (2):

$$E_{F_{(P)}} = E_{F_0}\left(\frac{m_{eff(0)}^{H}}{m_{eff(P)}^{H}}\right)\left(\frac{V_0}{V_P}\right)^{2/3} \tag{2}$$

In general, Eq. (2) provides estimates of E_F at ~93% or more of the DFT calculated value without corrections for charge redistribution along bond directions

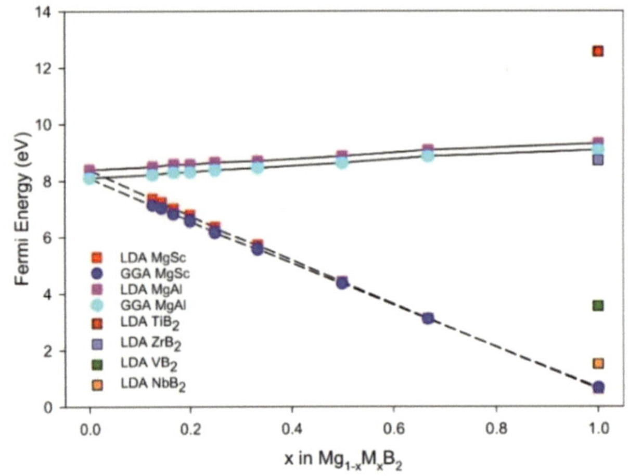

Figure 8.
Fermi energy (E_F) as function of metal substitution in $Mg_{1-x}Al_xB_2$ and $Mg_{1-x}Sc_xB_2$ calculated with the LDA and GGA functionals using the CASTEP module of materials studio for Δk = 0.02 Å^{-1}. Calculated Fermi energies for end-member compositions of AlB_2-type structures are also shown.

Pressure [GP$_a$]	Unit cell volume [A^3]	m_{eff}^{H} [m$_e$]*	Fermi energy [eV]	$E_{F(P)}$ [eV]	$\frac{E_{F-formula(1)}}{E_{F-DFT\ calculated}}$ [%]
0	27.8912	0.5070	8.4055	8.4055	100.0
2	27.5344	0.5059	8.5368	8.4964	99.5
3	26.5752	0.5035	8.9059	8.7411	98.1
8	25.7498	0.5018	9.2430	8.9572	96.9
9	25.0255	0.5003	9.5549	9.1565	95.8
14	24.3834	0.4990	9.8450	9.3409	94.9
20	23.8050	0.4983	10.1183	9.5049	94.0
38	23.2793	0.4981	10.3776	9.6513	93.0

$^*m_e$ is the electron mass.

Table 4.
List of calculated Fermi energy (E_F) values for MgB_2 at pressure using Eq. (2).

which take place as pressure is applied [55]. Use of parabolic approximations in this manner may provide a useful benchmark (or "rule of thumb") for models of predicted new compounds.

3.5 Phonon dispersions – variation with k-grid

As noted in earlier publications [17, 23, 25, 34], the k-grid value also influences the form and mode order of phonons in a DFT calculated PD. **Figure 6** demonstrates this influence on the MgB_2 PD for the range 0.02 Å^{-1} < Δk < 0.06 Å^{-1}. The more regularly shaped phonon anomaly becomes apparent with smaller k-grid value and is evident for Δk = 0.02 Å^{-1} (circled; **Figure 8a**). For values of Δk > 0.05 Å^{-1}, the calculated PD for MgB_2 implies that the phase is unstable yet we know from experimental evidence that this is not the case. For SC compounds with lower values of T_c and/or where Fermi surfaces closely intersect the Fermi level with minimal difference in reciprocal space, the sensitivity of the PD to k-grid value will be shifted towards smaller k-grids compared to the effect with MgB_2.

Measurement of the parameter, δ, shown in **Figure 9a** provides a reliable estimate of T_c for MgB_2 when the PD is calculated with $\Delta k < 0.02$ Å^{-1} [23]. This approach, which determines the thermal energy, T_δ, of the key E_{2g} phonon mode using an empirical formula [23], precisely tracks the experimentally determined reduction of T_c for metal-substituted forms of MgB_2 such as $(Mg_{1-x}Al_x)B_2$ and $(Mg_{1-x}Sc_x)B_2$ [23, 25]. When the value for δ is determined with two sequential DFT calculations using the LDA and the GGA functionals, error estimates (in terms of the amplitude of the spread of the DFT approximations) for the value of T_δ at each level of metal substitution can be obtained.

We have used this approach to estimate the likely value(s) of T_c for other metal-substituted forms of MgB_2 that have received limited attention or have not been identified previously in the literature. For example, we determined the PD for $(Mg_{1-x}Ba_x)B_2$, and for $(Mg_{1-x}Cd_x)B_2$ where x = 0.33, 0.5 or 0.66 [23, 25]. **Figure 10a** shows the PD for $(Mg_{0.5}Ba_{0.5})B_2$ calculated using the LDA functional with $\Delta k = 0.02$ Å^{-1}. Measurement of the four values for δ in **Figure 10a** (*i.e.* two non-degenerate E_{2g} modes each in the Γ–K and Γ–M directions) and conversion to T_δ gives an average value of 58.1 ± 3.4 K for $(Mg_{0.5}Ba_{0.5})B_2$. **Figure 10b** is adapted from the work of Palnichenko *et al.* [56] in which Ba, Rb and Cs were substituted into MgB_2 *via* solid state synthesis. In all cases, the T_c determined experimentally using magnetic susceptibility is higher than that for MgB_2. Unfortunately, while the effects of substitution are evident, the explicit levels of substitution were not determined [45].

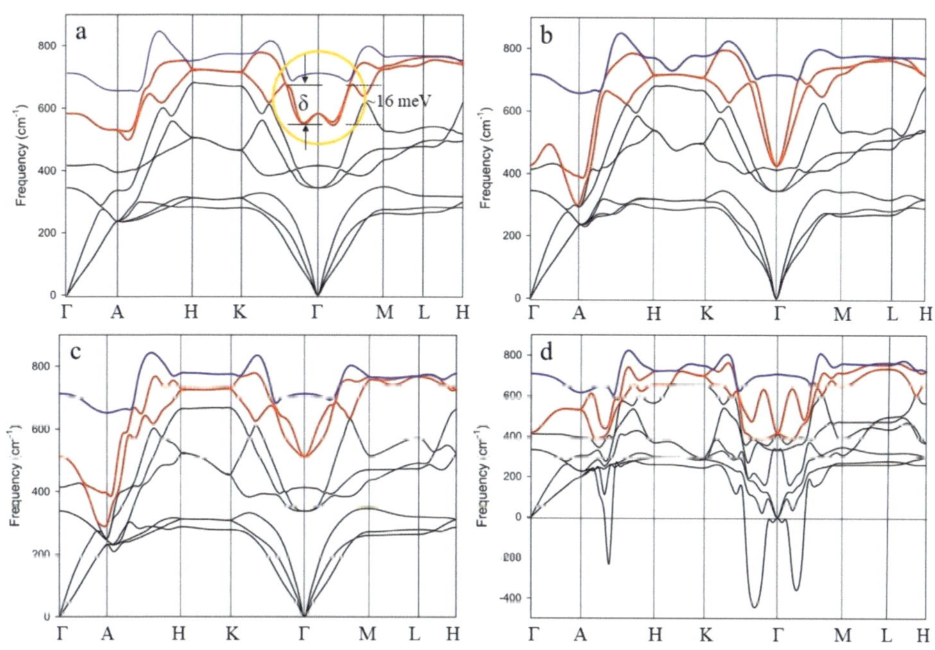

Figure 9.
Phonon dispersions for MgB_2 calculated using the LDA functional with different k-grids: (a) $\Delta k = 0.02$ Å^{-1}, (b) $\Delta k = 0.04$ Å^{-1}, (c) $\Delta k = 0.05$ Å^{-1} and (d) $\Delta k = 0.06$ Å^{-1}. E_{2g} (red) and B_{2g} (blue) phonon modes are highlighted. The energy associated with the E_{2g} phonon anomaly in (a) is ~16 meV [23]. Note the negative phonon frequencies for $\Delta k = 0.06$ Å^{-1} which, due to insufficient k-grid resolution, implies an unstable compound.

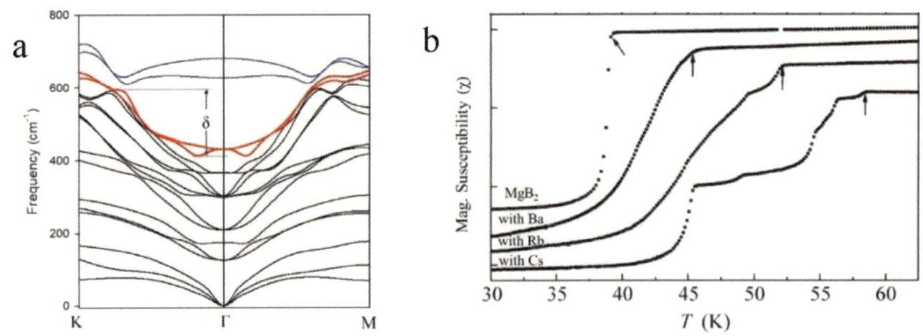

Figure 10.
(a) Phonon dispersion for $(Mg_{0.5}Ba_{0.5})B_2$ calculated using the LDA functional for Δk = 0.02 $Å^{-1}$, showing the extent, δ, of the E_{2g} phonon anomaly (in red) along the Γ–K and Γ–M directions; (b) magnetic susceptibility for MgB_2 and metal substituted forms of MgB_2 showing experimentally determined T_c values (arrowed); adapted from ***Figure 1*** *of Palnichenko et al. [56]. Substitution of Ba, Rb and Cs shows a higher T_c than for MgB_2.*

4. Discussion

The calculated E_F and Fermi level allow systematic comparison of EBSs from a structural family or group of materials with varying properties. Moreover, the FS is generally of a well-defined orbital character and topology determined by the value of E_F as a result of bands that cross the Fermi level. The Fermi level determined from DFT calculations is defined as at zero energy while the calculated value of E_F obtained after an accurate DFT calculation is seldom described in the published literature. By definition, the Fermi level is determined in the ground state by the filling of lower energy electronic states by all the (nearly) free electrons up to a highest possible value of the energy, which in practice should correspond to E_F [49, 50, 55, 57, 58].

4.1 Fermi energy

The superconducting gap in many compounds (*e.g.* diborides, disilicides, A15 compounds, B-doped diamond) is in the meV range of energy [59]. For many SC materials, the gap is directly linked to the separation of parallel, or nearly parallel, FSs that may not be identifiable if the k-grid value is at an insufficient resolution [45]. **Table 1** also shows that for MgB_2 differences in E_F and enthalpy of a few tens of meV are associated with exceptionally small differences in lattice parameter, of the order ~10^{-5} Å. This attribute highlights the robustness and sensitivity of DFT calculations, particularly when represented in reciprocal space. More importantly, these differences in E_F, attributed to differences in k-grid value, are substantially greater than the superconducting gap for MgB_2 [60]. Hence, detection of a gap – which in an EBS for MgB_2 is related to the separation of σ bands crossing the Fermi level – may not be achieved with low resolution DFT calculations.

We demonstrate this issue using the EBS for MgB_2 as shown in **Figure 11**. In this figure, we have reproduced the EBS for MgB_2 as calculated using the LDA functional for Δk = 0.018 $Å^{-1}$. The Fermi level is set at 0 eV and a notional "Fermi level 2" is also shown as a red dotted line at −250 meV. As noted in **Table 1**, a change in calculated E_F ~ 200 meV may occur with choice of Δk > 0.04 $Å^{-1}$. The intersection of σ bands with the calculated Fermi level are separated by a distance λ_1, which also defines the separation between Fermi surfaces for MgB_2.

However, if the value of Δk, or the calculated value for E_F, results in a shift of the Fermi level by ~250 meV, the separation of Fermi surfaces, illustrated by λ_2,

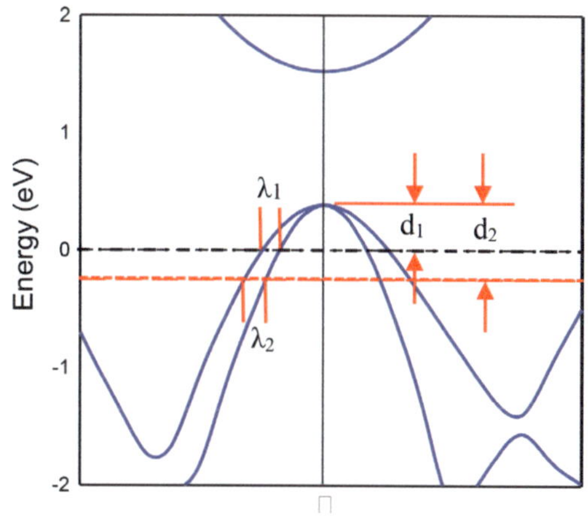

Figure 11.
Representative EBS showing the effect of a change in Fermi level of 250 meV (red dotted line). The values for λ_1 and λ_2, which define the distance between parallel Fermi surfaces, and the values for d_1 and d_2 (the energy above the Fermi level at the vertex of the σ band parabola), are not equivalent. This example is based on the EBS for MgB_2.

is different. We estimate that this order of Fermi level or E_F shift may result in discrepancies between 20% and 35% of the value(s) for λ. As shown in **Figure 11**, the shape of the σ bands around Γ are asymmetric. The difference in value(s) for λ with variation in E_F, will accordingly be dependent on the form, or shape, of these parabolas. Projections of the density of states at the Fermi level will also be affected by this shift of E_F as will the outcomes of Eliashberg or McMillan equations for the determination of T_c.

The apparent discrepancy in determination of the value for E_F, noted in the Introduction, may be elucidated by examination of **Figure 11**. For example, we suggest that some researchers define a value for E_F as equivalent to the energy shown as d_1 in **Figure 11** (*i.e.* the distance to the vertex of the parabolic band [61]). If this definition for E_F is used, then a shift in the Fermi level as described above would lead to an estimate equivalent to d_2 in **Figure 11**. The literature contains calculations of EBS where the specific approach to determine E_F is not identified; for superconductors, this practice provides an unfortunate level of uncertainty. An uncertain position for the Fermi level will also result in varying cross-sectional areas of the FS at the Fermi level (see **Figure 2**), which directly determines the period of sensitive quantum oscillation measurements [62]. Discrepancies between DFT predicted values and experimental quantum oscillations may be reconciled by revisiting the choice of k-grid.

4.2 Computational resolution

Table 5 provides a summary of reports on previous DFT calculations for MgB_2 and of systematic calculations from this study. This table highlights the diversity of computational methods used to date as well as wide variations in parameters such as k-grid value and the cut-off energy. Systematic evaluation of these two parameters shows that the value for E_F may differ by several hundred meV for the same cut-off energy with change in Δk value. For our systematic calculations of these parameters shown in **Table 5**, the LDA functional is used for consistency. Calculations with the GGA functional show similar trends albeit at different absolute values (by ~0.2 eV) for E_F.

Table 5 shows that a low value for cut-off energy (*i.e.* < 500 eV) results in a value for E_F > 1 eV different to that with cut-off energy >500 eV for calculations using the same Δk value (*e.g.* compare $\Delta k = 0.03$ Å^{-1} calculated for MgB_2 in **Table 5**). The importance of such parameters has been noted in the literature primarily in relation to PD calculations [17, 63]. However, the specific impact of both computational parameters on E_F and the effect on band structures has not previously been enumerated for MgB_2 nor for other SCs.

Table 5 also lists the variation in energy, ΔE_v (in eV), between the Fermi level and the vertex of the parabola at Γ for different values of Δk and for two cut-off energies using the LDA functional for the EBS of MgB_2 (in **Figure 11**, this energy is represented as d_1). As we have noted for E_F, there are substantial variations (*i.e.* > 100 meV) in ΔE_v with choice of Δk and cut-off energy. Calculated outcomes in our systematic study of MgB_2 parameters over a wide range of input parameters as listed in **Table 5**, show that for MgB_2, $\Delta k < 0.008$ Å^{-1} and a cut-off energy >900 eV, provides reliable determination of meV phenomena in this structure and in substitutional analogues of MgB_2. We note that these attributes apply to plane wave calculations. We are yet to undertake a systematic evaluation of augmented plane wave calculations using similar strategies.

The calculations by de la Pena-Seaman [71] on the transformation of Fermi surfaces with substitution of Al and C into MgB_2 and recent work by Pesic *et al.* [72] are notable exceptions on the previous studies shown in **Table 5** albeit each with a low cut-off energy. Note that a cut-off energy of 500 eV in **Table 5** results in Fermi energies similar to those obtained for molecular fragments obtained by the ADF software (data not shown). This suggests that calculations with smaller cut-off energy do not adequately capture periodic crystal behavior, but instead, model a set of values that are molecule-like. Some DFT studies reveal inherent inconsistencies in EBS and PD calculations for known superconductor materials due to insufficient computational resolution. This aspect of DFT models also appears to confuse the peer review process for some journal papers.

4.3 Phonon dispersions and k-grid

We have examined the changes in PD form and mode order for the substitutional series $Mg_{1-x}Al_xB_2$ [23] and $Mg_{1-x}Sc_xB_2$ [25] where $0 < x < 1$. For PDs, the value of k-grid in a DFT calculation may obscure phenomena that imply superconductivity such as the presence or absence of a phonon anomaly [5, 34]. We have also demonstrated for MgB_2 that the change in the E_{2g} phonon anomaly varies with applied pressure and correlates with the experimentally determined change in T_c [27]. For these cases, we show that a temperature, calculated from the extent of the anomaly, T_δ, is a reliable *ab initio* indicator of T_c determined by experiment [23, 24, 27]. A fine k-grid (or a k-grid value smaller than ~0.025 Å^{-1} depending on the structure) is important for PD plots of SCs with AlB_2-type structures and for estimations of T_c for BCS-type compounds that display a phonon anomaly [5, 23].

In an earlier publication [25], we compare for MgB_2 the calculation of T_c (i) using the McMillan formalism of the Eliashberg model [12] and (ii) using the E_{2g} phonon anomaly energy, T_δ, as noted above [23, 27, 28]. In both cases, with suitable assumptions for the McMillan formalism, the "predictive" fidelity of either method adequately matches experimental data. However, the Eliashberg model requires an estimate for two key parameters, λ and μ^*, based on average values of electron–phonon behavior summed over all orientations. In practice, determination of λ and/or μ^* by *a priori* methods is non-trivial for compounds with indeterminate physical properties [67]. In this regard, we applaud the recent advances in mathematical formalism and computational implementation of the Eliashberg model by Sanna *et al.* [19].

DFT Code	Δk value (Å^{-1})	Grid	No. of k-points	Energy Cut-off (eV)	Fermi Energy (eV)	$\Delta E_v{}^{\#}$ (eV)	LDA or GGA	Ref.
[64]	0.020	15x15x11	na	500	na	na	GGA	[38]
Eliashberg	*~0.022*	18x18x12	na	816	na	na	LDA	[65]
Eliashberg^{+}	*~0.018*	18x18x18	na	218	na	na	LDA	[17]
Eliashberg^{+}	*~0.050*	8x8x8	na	245	na	na	na	[66]
SCDFT	*~0.015*	24x24x22	na	340	na	na	GGA	[20]
VASP	*~0.019*	19x19x15	na	503	na	na	GGA	[67]
WIEN2k*	*~0.025*	14x14x14	144	122	na	na	GGA	[44]
WIEN2k	*~0.021*	18x18x14	na	82	na	na	LDA	[68]
SIESTA	*~0.018*	20x20x20	na	na	na	na	GGA	[69]
QExpr'o	*~0.015*	24x24x24	na	612	na	na	GGA	[70]
MBPP*	*~0.010*	36x36x36	na	218	na	na	Both	[71]
QExpr'o^{+}	*~0.007*	48x48x48	na	408	na	na	GGA	[72]
CASTEP	0.100	4x4x4	8	990	8.8103	−0.097	LDA	**
CASTEP	0.060	6x6x6	21	990	8.2110	0.636	LDA	**
CASTEP	0.050	8x8x6	30	990	8.5190	0.419	LDA	**
CASTEP	0.040	9x9x8	48	990	8.4706	0.328	LDA	[5]
CASTEP	0.030	12x12x10	95	990	8.3773	0.383	LDA	[5]
CASTEP	0.020	19x19x14	280	990	8.3976	0.348	LDA	[23]
CASTEP	0.015	25x25x20	650	990	8.3991	0.352	LDA	[5]
CASTEP	0.008	47x47x36	3744	990	8.4038	0.343	LDA	**

DFT Code	Δk value (Å^{-1})	Grid	No. of k-points	Energy Cut-off (eV)	Fermi Energy (eV)	ΔE_v[#] (eV)	LDA or GGA	Ref.
CASTEP	0.005	75x75x58	14703	990	8.4040	0.343	LDA	**
CASTEP	0.005	75x75x58	14703	500	7.3295	0.371	LDA	**
CASTEP	0.008	47x47x36	3744	500	7.3285	0.372	LDA	**
CASTEP	0.020	19x19x14	280	500	7.3218	0.370	LDA	**
CASTEP	0.030	12x12x10	95	500	7.2449	0.475	LDA	**

*"na" – data not available in publication. **This work; italicized values are estimates by this study.*
**Also describes (Mg,Al)B_2; grid size of EBS calculation is 18x18x18 for [71].*
[+]*Uses a 12x12x12 grid for PD calculations [72].*
[#]*Equivalent energy to d_1 in* **Figure 11**.

Table 5.
Comparison of computational settings for MgB_2.

An and Pickett [36] estimate that the influence of the E_{2g} mode is at least a factor of 25 times greater than all other phonon modes in MgB_2. The E_{2g} mode is predominantly associated with movement within the boron planes of MgB_2; that is, along specific orientations [73]. Nevertheless, use of an average value for phonon frequencies integrated over all directions in reciprocal space is a feature of the McMillan formalism that provides a reasonable "*post facto*" estimate of T_c presumably because the E_{2g} mode is so dominant. Such coincidence does not enable, nor guarantee, *ab initio* predictive capacity for *a priori* models, particularly if evaluating structures for which experimental data are limited or unavailable. Thus, we advocate an alternative approach for superconductivity prediction that complements the McMillan formalism. In this alternative approach, appropriate values for Δk and the cut-off energy enable *ab initio* DFT calculations to estimate values for T_δ that correlate with experimentally determined values for the T_c of MgB_2 [23, 27], for compounds of the form $(Mg_{1-x}M_x)B_2$ (where M = Al, Sc, Ti) [23–25], and for disilicides [23] and metal hexaborides [74].

The predictive value of the approaches we advocate to estimate T_c that utilizes calculation of a value for T_δ using a phonon anomaly [23–25, 74] is evident for Ba-substitution into MgB_2 [56]. Our estimates for $(Mg_{1-x}Ba_x)B_2$ at three levels of Ba substitution (x = 0.33, 0.5 and 0.66) and using both LDA and GGA approximations suggest that 62.1 K < T_δ < 64.4 K with an error of ±4.9 K. These estimates are higher by ~15 K than the experimentally determined value of ~45 K by Palnichenko *et al.* [56]. However, the extent of Ba substitution in MgB_2 was not determined in this experimental work; albeit ^{11}B NMR analysis shows that the final product has the same site symmetry as MgB_2 [56]. Substitution of Ba in MgB_2 at levels less than 33% may result in a lower value for T_c.

The presence of multiple phases in the Rb- and Cs- substituted forms of MgB_2 synthesized by Palnichenko *et al.* [56] is difficult to verify from the data presented due to limited microstructural and compositional characterization. However, we note that PD calculations on a nominal 50:50 ratio for Rb:Mg and Cs:Mg for substituted MgB_2 results in asymmetric and multi-level anomalies (data not shown) similar to that shown in **Figure 11a**. By measuring the extent of the anomaly in each of these cases, the values for T_δ are similar to the onsets of transitions shown for these compositions in **Figure 11b**. While circumstantial, this combination of modeling and experiment suggests that these substituted MgB_2 compositions may be homogeneous single phase. Further analyses of this compositional suite, and that of $(Mg_{1-x}Cd_x)B_2$ may reveal additional SC compounds in the AlB_2-type structural group with significantly enhanced superconducting properties to MgB_2.

Fully converged PDs are a useful indicator of phase stability [26, 74]. The sensitivity of PDs to changes in stoichiometry, composition or Δk is significantly higher than typically encountered in an EBS [26]. The PD calculated at a deliberately large k-grid value 0.06 $Å^{-1}$ in **Figure 9d** may be interpreted as a dynamic instability. MgB_2 is a well-studied case and we know that this is not correct; however, for unknown or other materials with closer FSs in reciprocal space, we would expect similar phenomena to be manifest at smaller k-grids. Thus, sometimes conclusions about phase transitions may be artifacts of the DFT calculation if k-grids of insufficient resolution are used for materials with approximately parallel FSs in close reciprocal space proximity [45].

4.4 Fermi surfaces and superconductivity

Electronic bands and FSs of constant energy possess all point symmetries of a crystal as a function of position in reciprocal space [48, 51]. The intersections of the σ bands with the Fermi level, as shown in **Figure 1**, determine points that, by construction, belong to the FSs. The FS corresponding to σ bands in the reduced BZ become two approximately parallel tubes [28], as schematically represented in **Figure 12** below. As shown for MgB_2 in earlier work [23, 24, 27, 28] and by others [65, 75, 76], these σ band FSs are not strictly cylindrical, but form as warped tubes with a narrowing in all directions towards Γ (sketched more accurately in **Figure 2**).

Since the FS tubes represent hole carrier sections, their interior will be empty in the ground state, while their exterior will be occupied. In a reduced zone schema, this construct creates ambiguous electron/hole character for the inter-tubular region. Ambiguity arises because this inter-tubular region should be, in the ground state, empty (*i.e.* without electrons) relative to the outer heavy effective mass σ band, but, in the ground state, filled with electrons relative to the inner light effective mass σ band.

This notion creates an apparent dilemma, although according to Ziman [55], "*There can be points in the zone where one cannot assign the label 'hole' or 'electron' uniquely to the states*". Further, "*the excitations of the superconducting state are peculiar quasi-particles which change from being 'electrons' to being 'holes' as they pass through the Fermi level*" [53]. Alternatively, we may reconcile this dilemma by considering that the reduced BZ scheme merges two different diameter tubes from points in reciprocal space within an extended BZ scheme [27, 28].

Given the indeterminate nature of the origin in reciprocal space, specific diameter tubes may be selected interchangeably by the DFT calculation; thus, implying a potential resonating behavior [28]. Analysis of electron–phonon behavior determined by DFT calculations suggests that this inter-tubular region of FSs (or other regions enclosed by parallel surfaces of different topology) is a region in reciprocal space that reveals the extent of superconductivity in typical BCS-type materials [5, 23, 24, 26–28]. Our calculations for both MgB_2 and B-doped diamond show that this inter-tubular region is of meV energy scale from the Fermi energy.

Parallel FSs are common features of superconducting compounds albeit their identification is dependent on crystal symmetry and the choice of k-grid value for DFT calculations [5, 23, 25]. The "resolution" of reciprocal space calculations using DFT (*i.e.* the value of k-grid) is a critical factor for identification of phenomena that may be influenced by changes of a few meV. For example, we show above that the value of E_F for MgB_2 may change by several hundred meV with a difference of ~0.02 $Å^{-1}$ for Δk (noting that T_c ~ 40 K). Such changes in E_F may shift the apparent Fermi level to a position where parallel FSs are not shown in an EBS. For compounds with a higher E_F value, closer to the parabola vertex (likely associated with lower T_c) and with larger difference in effective masses (*i.e.* the light mass displays a steeper EBS variation with k), the impact of this sensitivity to Δk increases.

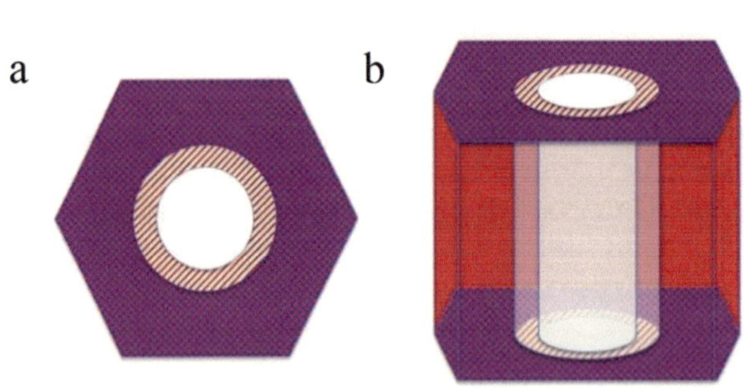

Figure 12.
Schematic of the FSs for MgB_2 viewed along: (a) the c-axis and (b) perpendicular to the c-axis. In this schematic, these FSs are simplified by neglecting warping in the DFT calculated model for MgB_2 [28]. Hatched section represents the inter-tubular region.

Thus, the value of k-grid used for DFT calculations is paramount. For PDs, this computational requirement has previously been well documented [14, 17, 20, 24] and, we suggest, is equally requisite for the use of EBS to predict, or design, new superconducting materials.

The recent development of an ML-based scheme to efficiently assimilate the function of the Kohn-Sham equation, and to directly and rapidly, predict the electronic structure of a material or a molecule, given its atomic configuration [41] is of salient interest with regard to k-grid value. This ML approach maps the atomic environment around a grid-point to the electron density and local density of states at that grid-point. The method clearly demonstrates more than two orders of magnitude improvement in computational time over conventional DFT calculations to generate accurate electronic structure details [41]. Utilization of this methodology at a k-point spacing <0.2 Å^{-1} to initialise ML-training for charge density [41] may enable very rapid determination of potential SC materials with many hundreds of atoms in the base structure. Nevertheless, as we have shown in this article, caution in the use of such values for Δk using ML is suggested because "false positives" for superconductivity may emerge and valid "hits" may be missed.

Thermal effects on electronic properties are generally included in DFT calculations as a smearing of electron behavior. However, high structural symmetry, or the lack of it, may impose significant anisotropy and/or preferred directionality of ionic movement that remains active even as temperature is increased. For reference, thermal excitation of the free-electron gas is k_BT or about 26 meV at ambient temperatures [51, 57]. As noted above, variations in E_F for superconducting phases may be in the meV range depending on the structure. We also note the importance of the smearing parameter in DFT calculations. We suggest that for particular superconducting cases where the T_c and/or phonon energy is low (*i.e.* $T_c < 10$ K) default values (~ 0.1–0.2 eV) in software packages for the smearing parameter may be misleading [77].

Calculated Fermi energies and Fermi levels are essential attributes for determination of materials properties in a range of other applications, such as for the energy band alignment of components in solar cell materials [78, 79], with solid-electrolyte interfaces [80], as well as for interface induced phenomena such as the substantial increase in T_c of monolayer FeSe on $SrTiO_3$ substrates [81]. Improved interpretation and understanding of electronic behavior in SCs and SC systems can be achieved with reliable calculated output values determined by *ab initio* DFT [82]. Indeed, Kohn posits that to achieve high accuracy with comprehensible representations of multi-particle systems, it is necessary to focus on real, three-dimensional coordinate space, *via* electron density distributions calculable using DFT [29].

5. Conclusions

The EBS encapsulates a wealth of information for superconductivity that may be misinterpreted due to the quality, or resolution, of DFT computations. A tendency to be satisfied with poor or limited computational resolution is evident in superconductivity literature unlike other fields that compute electronic properties using DFT. Translation of reciprocal space detail to real space periodicity for DFT-based design of new materials in an EBS with appropriate k-grid resolution can provide evidence for structures that may be viable SCs. As we have shown above, the E_F value is explicitly determined in DFT computations and, with consistent use of k-grid resolution, can provide comparable estimates of SC properties for proposed structures of a compositional suite. We encourage inclusion of these DFT calculated parameters in reports of SC materials.

We have described three fundamental approaches, based on *ab initio* DFT calculations to elucidate superconducting properties of existing and new compounds with relatively simple structures such as the AlB_2-type. This utilization of DFT, without modified functionals or estimates of free parameters, allows precise description of SC features in EBSs and PDs provided k-grid value and cut-off energy are optimized for high computational resolution. Through this process, we have identified a suite of AlB_2-type structures by metal substitution into MgB_2, that are likely to show higher T_c values than for MgB_2. These structures include compositions such as $(Mg_{1-x}M_x)B_2$ where M = Ba, Rb, Cs or Cd. In addition, the use of parabolic, or higher order quartic polynomials, to quantify key bands in an EBS offers a direct and low computational cost approach to determination of the superconducting gap for simple structures.

We are uncertain whether these approaches to DFT calculations apply to all SCs recognizing that now hundreds of compounds have been identified. Hardware and software limitations may restrict the use of these approaches to small unit cell structures of simple composition and higher symmetry. Nevertheless, in combination, these systematic and simple approaches to use of a well-known theory of electron distribution in solids suggest that prediction of properties for unknown, or hypothesized, SC structures is well within the reach of many materials researchers.

Acknowledgements

We appreciate discussions with Peter Talbot on molecular orbital calculations and DFT models. We also appreciate access to, and ongoing assistance with, QUT's HPC facilities, particularly from Hamish Macintosh, Abdul Sharif and Ashley Wright of the e-Research office. This research did not receive any specific grant from funding agencies in the public, commercial or not-for-profit sectors. AA is grateful for generous financial support for higher degree studies from the University of Hafr AlBatin, Saudi Arabia.

All data referred to in and underpinning this publication are available in QUT Research Data Finder and can be found at DOI: 10.25912/5c8b2cc59a2d9.

Conflict of interest

All authors declare that they have no conflict of interest.

Author details

Ian D.R. Mackinnon[1*], Alanoud Almutairi[1,2] and Jose A. Alarco[1]

1 Institute for Future Environments and Science and Engineering Faculty, Queensland University of Technology (QUT), Brisbane, QLD, Australia

2 University of Hafr AlBatin, Hafar Al Batin, Saudi Arabia

*Address all correspondence to: ian.mackinnon@qut.edu.au

References

[1] Sutton AP. Electronic Structure of Materials. Oxford UK Clarendon Press: Oxford Science Publications 2004.

[2] Phillips JC. Bonds and Bands in Semiconductors: Academic Press 1973.

[3] Hoffmann R. How Chemistry and Physics Meet in the Solid State. Angew Chem Int Ed. 1987;26(9):846-878.

[4] Berger RF, Walters PL, Lee S, Hoffmann R. Connecting the Chemical and Physical Viewpoints of What Determines Structure: From 1-D Chains to γ-Brasses. Chem Rev. 2011;111:4522-4545.

[5] Alarco JA, Chou A, Talbot PC, Mackinnon IDR. Phonon Modes of MgB_2: Super-lattice Structures and Spectral Response. Phys Chem Chem Phys. 2014;16:24443-24456.

[6] Uchida S-i. Overview of Superconducting Materials with T_c Higher than 23 K. High Temperature Superconductivity. Japan: Springer 2015. p. 7-21.

[7] Alexandrov AS. Nonadiabatic polaronic superconductivity in MgB_2 and cuprates. Physica C. 2001;363:231-236.

[8] Malik GP. On the role of fermi energy in determining properties of superconductors: A detailed comparative study of two elemental superconductors (Sn and Pb), a non-cuprate (MgB_2) and three cuprates (YBCO, Bi-2212 and Tl-2212). J Supercond Nov Magn. 2016;29:2755-2764.

[9] Clark SJ, Segall MD, Pickard CJ, Hasnip PJ, Probert MIJ, Refson K, et al. First principles methods using CASTEP. Z Kristallogr. 2005;220:567-570.

[10] Refson K, Tulip PR, Clark SJ. Variational density-functional perturbation theory for dielectrics and lattice dynamics. Phys Rev B. 2006;73(155114):1-12.

[11] ADF Manual 2018 Available from: www.scm.com.

[12] Liu AY, Mazin II, Kortus J. Beyond Eliashberg Superconductivity in MgB_2: Anharmonicity, Two-Phonon Scattering, and Multiple Gaps. Phys Rev Lett. 2001;87:087005.

[13] Carbotte JP. Properties of boson exchange superconductors. Rev Modern Phys. 1990;62(4):1027-1157.

[14] Kortus J. Current progress in the theoretical understanding of MgB_2. Physica C. 2007;456:54-62.

[15] Joon Choi HJ, Louie SG, Cohen ML. Prediction of super conducting properties of CaB_2 using anisotropic Eliashberg theory. Phys Rev B. 2009;80(064503):1-4.

[16] Zheng J-C, Zhu Y. Searching for a higher superconducting transition temperature in strained MgB_2. Phys Rev B. 2006;73(2):024509.

[17] Heid R, Bohnen K-P, Renker B. Electron-phonon coupling and superconductivity in MgB_2 and related diborides. Adv Solid State Phys. 2002;42:293-305.

[18] d'Astuto M, Heid R, Renker B, Weber F, Schober H, De la Pena-Seaman O, et al. Nonadiabatic effects in the phonon dispersion of $Mg_{1-x}Al_xB_2$. Phys Rev B. 2016;93(180508R):1-5.

[19] Sanna A, Flores-Livas JA, Davydov A, Profeta G, Dewhurst K, Sharma S, et al. *Ab initio* Eliashberg Theory: Making Genuine Predictions of Superconducting Features. Journal of the Physical Society of Japan. 2018;87(041012):1-8.

[20] Floris A, Sanna A, Luders M, Profeta G, Lathiotakis NN, Marques MAL, et al. Superconducting properties of MgB_2 from first principles. Physica C. 2007;456:45-53.

[21] Oliveira LN, Gross EKU, Kohn W. Phys Rev Lett. 1988;60:2430.

[22] Drozdov AP, Eremets MI, Troyan IA, Ksenofontov V, Shylin SI. Conventional superconductivity at 203 kelvin at high pressures in the sulfur hydride system. Nature. 2015;525:73-76.

[23] Alarco JA, Talbot PC, Mackinnon IDR. Phonon anomalies predict superconducting T_c for AlB_2-type structures. Phys Chem Chem Phys. 2015;17(38):25090-25099.

[24] Alarco JA, Talbot PC, Mackinnon IDR. A complete and accurate description of superconductivity of AlB_2-type structures from phonon dispersion calculations. J Supercond Novel Magnet. 2018;31:727-731.

[25] Mackinnon IDR, Talbot PC, Alarco JA. Phonon dispersion anomalies and superconductivity in metal substituted MgB_2. Comp Mater Sci. 2017;130:191-203.

[26] Alarco JA, Mackinnon IDR. Phonon dispersions as indicators of dynamic symmetry reduction in superconductors. In: Stavrou VN, editor. Phonons in Low Dimensional Structures. London UK: InTech Open; 2018. p. 75-101.

[27] Alarco JA, Talbot PC, Mackinnon IDR. Phonon dispersion models for MgB_2 with application of pressure. Physica C: Supercond and Applications. 2017;536:11-17.

[28] Alarco JA, Talbot PC, Mackinnon IDR. Electron density response to phonon dynamics in MgB_2: an indicator of superconducting properties. Mod Numer Sim Mater Sci. 2018;8:21-46.

[29] Kohn W. Nobel Lecture: Electronic structure of matter—wave functions and density functionals. Rev Modern Phys. 1999;71(5):1253-1266.

[30] Kohn W. Fundamentals of density functional theory. Density Functionals: Theory and Applications. 1998;500:1-7.

[31] Lee KW, Pickett WE. Boron spectral density and disorder broadening in B-doped diamond. Phys Rev B. 2006;73(7):075105.

[32] Lee KW, Pickett WE. Superconductivity in Boron-Doped Diamond. Phys Rev Lett. 2004;93(23):237003.

[33] Tomioka F, Tsuda S, Yamaguchi T, Kawarada H, Takano Y. Pressure effect of superconducting transition temperature for boron-doped diamond films. Physica C. 2008;468:1228-1230.

[34] Baron AQR, Uchiyama H, Tanaka Y, Tsutsui S, Ishikawa D, Lee S, et al. Kohn Anomaly in MgB_2 by Inelastic X-Ray Scattering. Phys Rev Lett. 2004;92:197004.

[35] Bohnen K-P, Heid R, Renker B. Phonon Dispersion and Electron-Phonon Coupling in MgB_2 and AlB_2. Phys Rev Lett. 2001;86(25):5771-5774.

[36] An JM, Pickett WE. Superconductivity of MgB_2: Covalent Bonds Driven Metallic. Phys Rev Lett. 2001;86(19):4366-4369.

[37] Kong Y, Dolgov OV, Jepsen O, Andersen OK. Electron-phonon interaction in the normal and superconducting states of MgB_2. Phys Rev B. 2001;64(020501):1-5.

[38] Yildirim T, Gülseren O, Lynn JW, Brown CM, Udovic TJ, Huang Q, et al.

Giant Anharmonicity and Nonlinear Electron-Phonon Coupling in MgB_2: A Combined First-Principles Calculation and Neutron Scattering Study. Phys Rev Lett. 2001;87(3):037001.

[39] Kunc K, Loa I, Syassen K, Kremer R, Ahn K. MgB_2 under pressure: phonon calculations, Raman spectroscopy, and optical reflectance. J Phys: Condens Matter. 2001;13(44):9945-9962.

[40] Baron AQR, Uchiyama H, Heid R, Bohnen KP, Tanaka Y, Tsutsui S, et al. Two-phonon contributions to the inelastic x-ray scattering spectra of MgB_2. Phys Rev B. 2007;75:020505.

[41] Chandrasekaran A, Kamal D, Batra R, Kim C, Chen L, Ramprasad R. Solving the electronic structure problem with machine learning. Comp Mater. 2019;5(22):1-7.

[42] Alarco JA, Talbot PC, Mackinnon IDR. Phonon anomalies predict superconducting T_c for AlB_2-type structures. Physical Chemistry Chemical Physics. 2015;17(38):25090-25099.

[43] Sherwood PMA. Vibrational Spectroscopy of Solids. Cambridge UK: Cambridge University Press 1972.

[44] de la Penha O, Aguayo A, de Coss R. Effects of Al doping on the structural and electronic properties of $Mg_{1-x}Al_xB_2$. Phys Rev B. 2002;66:01251.

[45] Alarco JA, Almutairi A, Mackinnon IDR. Progress towards a universal approach for prediction of the superconducting transition temperature. J Supercond Nov Magn. 2020;33:2287-2293.

[46] Nagamatsu J, Nakagawa N, Muranaka T, Zenitani Y, Akimitsu J. Superconductivity at 39 K in magnesium diboride. Nature. 2001;410:63-64.

[47] Harrison WA. Electronic structure and the properties of solids: the physics of the chemical bond. New York: Dover Publications, Inc.; 2012.

[48] Jones H. The Theory of Brillouin Zones and Electronic States on Crystals. Amsterdam: North-Holland Publishing Company 1960.

[49] Ashcroft NW, Mermin ND. Solid State Physics. Saunders, Philadelphia. 1976;293.

[50] Singleton J. Band theory and electronic properties of solids: Oxford University Press 2001.

[51] Altmann SL. Band Theory of Solids: An Introduction from the Point of View of Symmetry: Clarendon Press, Oxford 2002.

[52] Tinkham M. Group Theory and Quantum Mechanics. New York: Dover Publications, Inc.; 2003.

[53] Ziman JM. Principles of the Theory of Solids. Second ed: Cambridge University Press 1972.

[54] Buzea C, Yamashita T. TOPICAL REVIEW- Review of the superconducting properties of MgB_2. Supercond Sci Technol. 2001;14:R115–RR46.

[55] Ziman JM. Electrons and Phonons: The Theory of Transport Phenomena in Solids: Oxford University Press 1960.

[56] Palnichenko AV, Vyaselev OM, Sidorov NS. Influence of Rb, Cs, and Ba on the Superconductivity of Magnesium Diboride. JETP Lett. 2007;86(4):272-274.

[57] Duan F, Guojun J. Introduction to Condensed Matter Physics World Scientific 2007.

[58] Mizutani U. Introduction to the Electron Theory of Metals. Cambridge Cambridge University Press; 2001.

[59] Poole CPJ, Farach HA, Creswick RJ, Prozorov R. Super conductivity. Burlington Elsevier Science 2014.

[60] Hott R, Kleiner R, Wolf T, Zwicknagl G. Review on Superconducting Materials. Encyclopedia of Applied Physics Wiley-VCH 2019. p. 60.

[61] Boeri L, Bachelet GB, Cappelluti E, Pietronero L. Small Fermi energy and phonon anharmonicity in MgB_2 and related compounds. Phys Rev B. 2002;65(21):214501.

[62] Rhodes D, Schönemann R, Aryal N, Zhou Q, Zhang QR, Kampert E, et al. Bulk Fermi surface of the Weyl type-II semimetallic candidate γ -$MoTe_2$. Phys Rev B. 2017;96(165134):1-12.

[63] Calandra M, Mauri F. Electron-phonon coupling and phonon self-energy in MgB_2: Interpretation of MgB_2 Raman spectra. Phys Rev B. 2005;71(064501):1-10.

[64] Payne MC, Teter MP, Allan DC, Arias TA, Joanopoulous JD. Iterative minimization techniques for *ab initio* total-energy calculations: molecular dynamics and conjugate gradients. Rev Modern Phys. 1992;64(4):1045-1097.

[65] Choi HJ, Roundy D, Sun H, Cohen ML, Louie SG. The origin of the anomalous superconducting properties of MgB_2. Nature. 2002;418:758-760.

[66] Chen X-R, Guo H-Z, Cai L-C, Gao J. First principles calculation for the electron-phonon coupling constant of superconducting MgB_2. Chin Phys Lett. 2005;22(6):1504-1506.

[67] Neaton JB, Perali A. On the possibility of superconductivity at higher temperatures in sp-valent diborides 2001. Available from: http://arxiv.org/abs/cond-mat/0104098v1.

[68] Benhelal O, Laksari S, Chahed A, Khatir R. Full-potential study of structural and electronic properties of MB_2-type metal diborides (M = Be, Mg and Ca). Comp Mater Sci. 2011;50:1889-1893.

[69] Sharma D, J. Kumar, A. Vajpayee, R. Kumar, P.K. Ahluwalia, Awana VPS. Comparative experimental and density functional theory (DFT) study of the physical properties of MgB_2 and AlB_2. J Supercond Nov Magn. 2011;24:1925-1931.

[70] Morshedloo T, Roknabadi MR, Behdani M. First-principles study of the superconductivity in MgB_2 bulk and in its bilayer thin film based on electron–phonon coupling. Physica C. 2015;509:1-4.

[71] De la Peña-Seaman O, de Coss R, Heid R, Bohnen K-P. Effects of Al and C doping on the electronic structure and phonon renormalization in MgB_2. Phys Rev B. 2009;79(134523):1-8.

[72] Pesic J, Popov I, Šolaji A, Damljanovic V, Hingerl K, Belic M, et al. Ab Initio Study of the Electronic, Vibrational, and Mechanical Properties of the Magnesium Diboride Monolayer. Condens Matter. 2019;4:1-10.

[73] Kunc K, Loa I, Syassen K, Kremer RK, Ahn K. MgB_2 under pressure: phonon calculations, Raman spectroscopy, and optical reflectance. J Phys: Condens Matter. 2001;13:9945-9962.

[74] Alarco JA, Shahbazi M, Talbot PC, Mackinnon IDR. Spectroscopy of metal hexaborides: Phonon dispersion models. J Raman Spect 2018;49:1985-1998.

[75] Mazin II, Antropov VP. Electronic structure, electron–phonon coupling, and multiband effects in MgB_2. Physica C: Superconductivity. 2003;385(1):49-65.

[76] Kortus J, Mazin II, Belashchenko KD, Antropov VP, Boyer LL. Superconductivity of Metallic Boron in MgB_2. Phys Rev Lett. 2001;86(20):4656-4659.

[77] Alarco JA, Talbot PC, Mackinnon IDR. Comparison of functionals for metal hexaboride band structure calculations. Mod Numer Sim Mater Sci. 2014;4:53-69.

[78] Klein A, Korber C, Wachau A, Sauberlich F, Gassenbauer Y, Harvey SP, et al. Transparent Conducting Oxides for Photovoltaics: Manipulation of Fermi Level,Work Function and Energy Band Alignment. Materials. 2010;3:4892-4914.

[79] Aygüler MF, Hufnagel AG, Rieder P, Wussler M, Jaegermann W, Bein T, et al. Influence of Fermi Level Alignment with Tin Oxide on the Hysteresis of Perovskite Solar Cells. ACS Applied Materials and Interfaces. 2018;10:11414-11419.

[80] Bockris JOM, Reddy AKN. Modern Electrochemistry - An Introduction to an Interdisciplinary Area. New York: Plenum Plubishing Corporation; 1973.

[81] Ge J-F, Liu Z-L, Liu C, Gao C-L, Qian D, Xue Q-K, et al. Superconductivity above 100 K in single-layer FeSe films on doped $SrTiO_3$. Nature Materials. 2014;14:285-289.

[82] Cohen AJ, Mori-Sanchez P, Yang W. Challenges for Density Functional Theory. Chem Rev. 2012;112:289-320.